The Impact Challenge

Impactful Data Science

Series Editor
Global Association for Research Methods and Data

The Impact Challenge
Reframing Sustainability for Businesses
Alessia Falsarone

For more information about this series, please visit: www.routledge.com/Impactful-Data-Science/book-series/IDS

The Impact Challenge
Reframing Sustainability
for Businesses

Alessia Falsarone

CRC Press
Taylor & Francis Group
Boca Raton London New York

CRC Press is an imprint of the
Taylor & Francis Group, an **informa** business

Cover image: The image was created and provided by the author Alessia Falsarone

First edition published 2022
by CRC Press
6000 Broken Sound Parkway NW, Suite 300, Boca Raton, FL 33487-2742

and by CRC Press
4 Park Square, Milton Park, Abingdon, Oxon, OX14 4RN

CRC Press is an imprint of Taylor & Francis Group, LLC

© 2022 Taylor & Francis Group, LLC

Library of Congress Cataloging-in-Publication Data
Names: Falsarone, Alessia, author.
Title: The impact challenge : reframing sustainability for businesses /
Alessia Falsarone.
Description: Boca Raton, FL : CRC Press, 2022. | Includes bibliographical
references and index.
Identifiers: LCCN 2021049264
Subjects: LCSH: Organizational behavior. | Business enterprises—Data
processing. | Sustainable development.
Classification: LCC HD58.7 .F34 2022 | DDC 302.3/5—dc23/eng/20220112
LC record available at https://lccn.loc.gov/2021049264

ISBN: 9781032079509 (hbk)
ISBN: 9781032079516 (pbk)
ISBN: 9781003212225 (ebk)

DOI: 10.1201/9781003212225

Typeset in Times New Roman
by codeMantra

Contents

Acknowledgments

The frameworks, tools and concepts presented in this book would not be possible without the collaboration, encouragement, advice and suggestions from many colleagues across sectors and academic fields. Over the years, they have provided an exceptional sounding board of ideas in countless ways. Most importantly, they shared the journey and provided significant inspiration to the nascent body of work that drives many impactful business transitions. Among such powerful voices, I am particularly thankful for the milestones that the following group (listed in no particular order) is setting in advancing the practice of impact to business and financial management: Aurelio Bustilho de Oliveira and Rafael de la Haza Casarrubio (Enel Americas); Ryohei Yanagi, PhD (Eisai Co., Ltd. and Waseda University); Rick Plympton (Optimax Systems); Stuart Coleman (The Open Data Institute); Anthony Cowell and Niven Huang (KPMG); Suzette Carty (Brown-Forman); Anjuli Pandit (HSBC); Anton Gorodniuk (Wells Fargo Asset Management); Bob Hirth (Protiviti); Janine Guillot and Katie Schmitz Eulitt (The Value Reporting Foundation); Marvin Smith (Gap Inc.), Eduardo Alfonso Atehortua Barrero, Chris Fowle and Carmen Nuzzo (Principles for Responsible Investment); Philipp Aeby, Danilo Chiono and Alexandra Mihailescu Cichon (RepRisk AG); Marie-Josée Privyk (Novisto); Esohe Denise Odaro (International Finance Corporation); Eila Kreivi (European Investment Bank); Soffia Alarcon (IHS Markit and Carbon Trust); Pablo Lumerman (Grupo de Diálogo Norpatagónico); Mathieu Verougstraete (United Nations); Amr Addas (Concordia University); Lutz Kilian (Federal Reserve Bank of Dallas); Johanna Hising DiFabio, Retsef Levi and Stewart Myers (MIT Sloan School of Management); Nathan Williams (Minespider); Faith Ward (Brunel Pension); My-Linh Ngo (BlueBay Asset Management); Jennifer Jordan (Techstars); Fernando Dangond (EMD Serono); Johanna Köb (Zurich Insurance); Mariuz Calvet (Grupo Financiero Banorte); Mauricio Rodriguez (Protección SA); Tsewang Namgyal (Mitsubishi UFJ Financial Group); Howard Coon (Avangrid); Mary C. Gentile, PhD. (UVA Darden School of Business); Rahul Raj (5&Vine); Suzanne Gibbs Howard (IDEO U); Judy Samuelson, Nancy McGaw, Eli Malinsky and Rachel Botos (The Aspen Institute); Sonja Haut (Novartis); Andrew Ohm (Starbucks Coffee Company); Bonnie Lei and Jaime Galviz (Microsoft); James Ossman (Etsy); Carolina García Arbeláez (AB InBev); Zack Langway (Johnson & Johnson); Jason A. Scott (Google); Lindsey Blumenthal (Apple); Kelsea Ballantyne (The Boeing Company); Jennifer Michael (Chevron); Timothy Howe (Cox Enterprises); Kaity M. Ruger (CVS Health); Joan Bohan (The Walt Disney Company); Susie Nam (Accenture Interactive | Droga5); Chaya Nayak (Meta Platforms); Megan Weibler (IDEO); Larcy Cooper (Paul, Weiss, Rifkind, Wharton & Garrison LLP); Nicole Horvath (Pfizer); Jack Soos (Pratt & Whitney); Patrick Liang (Tang Industries); Emily Alati (Vans); Mitchell Roshong (Institute of Management Accountants); Andreas Simou (Center for Financial Professionals CeFPro®).

In addition to the industry colleagues and professionals listed above, I also wish to thank the following family members and friends. Without their inspiration, encouragement and endless support, this book would never have come to fruition.

To Martha, for inspiring me every day to look at the world through the eyes of the new generation and generations to come.

To Rob, for challenging me to only work on things that matter and with people that care as much as I do.

To Anna and Antonio, because my inner resilience, deep-rooted integrity and desire to work hard for what I believe in all come from you.

To Danalee, Luca and Mike, for your immense reservoir of patience and your witty sense of humor.

To the Fellows of the Aspen Institute Business and Society Program, the Unmutables, for making 2020 my year of positive change.

To the 2021 Women inPower fellows at the 92Y Belfer Center for Social Impact and Innovation and my mentor through the program, Mai-Ahn Tran (The Ford Foundation), for always raising leaders to the top.

About the Author

Alessia Falsarone, SASB FSA, is a sustainable finance expert and a fellow of the Aspen Institute Business and Society Program. Her work bridges the gap between sustainability, financial innovation and risk management. A sought-after commentator for media outlets and contributor to academic programs, Ms. Falsarone is a member of high-level advisory groups that promote environmental and climate finance, including the G20 Environmental Ministerial, the London Stock Exchange, the Sustainability Accounting Standards Board (Value Reporting Foundation) and the UN Principles for Responsible Investment. In recognition of her innovative vision for business and society, she has received an Honoree Award from the Women's Venture Fund and the 2021 Global Leadership Award by the SheInspires Foundation in the UK. She is an alumna of Stanford University, the MIT Sloan School of Business and Bocconi University. Ms. Falsarone holds certified director status with the National Association of Corporate Directors. An avid advocate of sustainability in business education, she has contributed to educational initiatives on the topic at the Asian University for Women, the Society of Corporate Compliance, the Swiss Sustainable Finance Initiative, the United Nations, the World Bank, Stanford University and University of Chicago, including delivering training on climate risk and green finance in Asia Pacific and Latin America.

Introduction

We live in the decade of ESG: Environmental and Social Governance. Rather than being one item in a list of practices needing attention as we sail further into the twenty-first century, it is governance – that set of tools, policies, behaviors and organizational beliefs – that must take the helm and guide business and individual actions to define a path to generate economic and societal values.

Organizations are being forced into a leadership introspection. Whether they are large international corporations or small, domestic players, and whether they're willing or not, businesses are on a journey to rethink how they deliver value, to whom and according to which principles. Above all, they are challenged to define a new path for organizational incentives and decision-making powered by new sources of data. Whether tasked to address the impact of water scarcity, the provenance of goods through their supply chains, issues around human rights abuses or the sustenance of indigenous populations, leaders at all levels are struggling to incorporate ESG criteria in an intentional and systematic fashion in the way they deliver business outcomes while also creating a new learning roadmap that sits at the core of the companies they are part of.

The "impact challenge" is to turn to sustainability as a society by developing ESG operating frameworks, one business at a time. The challenge posed by the massive amount of data emerging from early attempts to evaluate ESG practices is that most data sources and available evaluation systems remain outside an organization's range of knowledge. Ironically, the first attempts at integrating ESG dimensions into the fabric of businesses are relegating these variables to external knowledge hubs, creating a multitude of *data technicians* instead of *business strategists*. This is not surprising, as companies are seeking to balance their existing operating context – what they know and already have access to – with a fast-evolving regulatory environment that seems to be creating more and more minimum requirements to comply with.

We need to think deeply about how we want to reshape organizations to learn about sustainability, in what terms and why. We must ask: Who is responsible and how do we ensure the inevitable learning is long lasting? I have spent the past two decades witnessing how three simple dimensions can help analyze how closely intertwined societal and financial issues are in domestic and international markets. I define those dimensions as Context, Impact and Value. Organizations that redefine their sustainability journey starting from their individual ecosystem (their Context) and the mission they are willing to deliver (their Impact) are best positioned to truly learn from the wealth of insights that stem from the emergence of multidisciplinary approaches to translate sustainability principles into actionable organizational roadmaps. Those businesses are the ones that recognize and prioritize ESG as a business responsibility and as a source of enterprise value (Value).

DOI: 10.1201/9781003212225-1

1

In the post-COVID world, in which we face a much needed societal and economic recovery, these pillars are even more compelling. We are all, somehow, some way, on a path to building a more satisfying, engaging, purposeful reality through the alignment of values and value – for us, our loved ones and future generations. And we need science to take the driver's seat to get us to the place where we are struggling to get to – struggling because anxieties, societal challenges and malaise have affected our decision-making. COVID has hit us in many organizational contexts – in the private and public sector, at the neighborhood and household level – and we are suffering through the crisis in big ways. Issues such as mental health, disrupted work–balance and the future of wealth transfer – to women and minorities – have been elevated to major roadblocks to a sustained recovery.

We are rebuilding and need to rebuild trust in the way decisions are enforced. We must learn together, not alone. The repercussions of leaving leadership at the top of the funnel means that incentives schemes may not be properly aligned to a common purpose, and the impact challenge may take much longer to solve.

This book is written for those seeking to adopt the pillars of sustainability innovation as a core part of strategic business growth. It starts with the raw data and ends with a programmatic and intentional adoption of new sources of insights. It starts and ends with driving organizational learning toward impact objectives as positive societal changes. The businesses that need it most are those that define sustainability commitments – whether environmental pledges to a zero-emission future or diversity targets to achieve equitable representation – through a *cost of doing business* mindset without exploring the potential for socio-economic and environmental wealth created by organizational learning. In this regard, the role of academics and practitioners in promoting sustainability literacy is pivotal.

There is lot of data and product innovation to meet sustainability features, goals and aspirations on the back of the rise of ESG investors. But there is not as much process innovation yet for businesses to adapt and adopt sustainability criteria and build a sustainable DNA within their organizations.

A note to my audience: The leaders – those who recognize the impact challenge – are sitting at the top, in the middle and at the bottom of the organizational funnel. You all matter equally to shape the ESG decade. Your intuition, and the way you see the world and rebuild through your choices, is key. We are in the era of big, bold sustainability pledges brought forward by businesses. Which ones will be delivered? How do we bridge the gap between organizational behaviors and long-range commitments? This book will deliver the tools to inquire, think through and realign organizational frameworks in a programmatic and intentional manner.

The first five chapters equip you with frameworks to map the impact journey that your organization faces and highlight the role of data in making the learning experience of teams a highly rewarding one. Inspired by the direct experience of practitioners that have started the process earlier, the second half of the book will look at the future, at what lies beyond pledges and how each reader can contribute directly to make the impact challenge an opportunity. At the end of each chapter, the Learning Journey sections will help advance the analyses, data methods and multidisciplinary frameworks highlighted to the case of an organization of choice. In addition, the Technical Notes bring forward developments in data science for impact.

My sincere hope is that this publication will provide value to students entering the world of business as well as seasoned professionals. That beyond building a functional expertise in sustainability practices applicable to your immediate role, you may be able to bring principles of impact to the analytical realm of data. This may be in institutions already on the sustainability journey as well as those that will emerge as a result of structural failures.

To my colleagues across sectors and geographies who have shared that journey with me over the years, I am thankful for your decision to show up every day and contribute to build stronger institutions. Keep inspiring. It does not go unnoticed. Your contributions are shaping the future of data-driven organizations led through purposeful decision-making, to a more resilient and intentional future. You are leading beyond functional titles and political appointments to unbiased, unequivocally just and science-based impact goals built on trust, a spirit of organizational inquiry and impactful learning. This publication speaks of your work.

1 The Impact Enterprise

Over the past twenty years, sustainability efforts in the private sector have lacked the organizational structures they need to create long-lasting results. This has made it difficult to replicate or scale them outside of an individual market, product or team. A new field of study – that of data science for impact – promises to address this challenge and reshape the way organizations learn and act on impact-oriented business targets that combine the pursuit of positive societal outcomes with traditional financial metrics over a sustained period of time. What is data science for impact? I define it as a multidisciplinary set of research methods and analytical processes that can help an enterprise to achieve its impact objectives. Crucially, it must also help the enterprise define its own organizational learning path to sustain those objectives.

The challenges tackled by enterprises that commit to impact targets are multifaceted – from climate change and preservation of natural ecosystems to the need for inclusivity, equanimity and diversity. This means that in developing roadmaps and solutions, organizations that seek to deliver upon those targets must build a spectrum of collective efforts that foster authentic dialogue and constructive exchange of ideas between today's scientists and industry practitioners and the future generation of scientists and industry practitioners they influence. For the purpose of this publication, I will use the terms *outcomes*, *impact* and *purpose* interchangeably.

THE EVOLUTION OF DATA SCIENCE FOR IMPACT

Historically, the definition of *impact* has been linked to the development of evaluation and assessment tools that are applied in environmental and ecological studies or in the area of socio-economic development. In the latter case, this was mostly led by international organizations and not-for-profit institutions as a way to value change and support the case for development funding – such tools were used to measure improvement in a variety of indicators of human wellbeing. The investigation of impact assessment tools dates back to the 1950s as an attempt by international NGOs and development agencies to forecast the likely environmental, social and economic consequences of carrying out a specific program. Environmental and social cost-benefit analyses (CBAs) were integrated into their decision-making. Generally, the effectiveness of CBAs was evaluated when a project or initiative was completed. This backward-looking view helped address a core *resource allocation problem* for their program portfolios (as a linear programming exercise) and identify whether the program was allocated enough people and funding to succeed.

Longitudinal approaches were introduced shortly thereafter to complement the simple resource allocation exercise of CBAs. In the evolution of the impact

DOI: 10.1201/9781003212225-2

assessment toolkit, longitudinal studies added a component of forward-looking assessment. They helped track progress in target areas of impact at regular intervals, not just at the completion of a project. This moved efforts a step closer to leveraging direct evidence from pilot projects to maximize the target impact while keeping resource allocation constraints in mind. Identifying the set of descriptive variables that define the impact objective constitutes the core element of socio-economic studies that employ a longitudinal approach, while also making them comparable among a portfolio of potential impact projects an organization can choose from.

Let us take the example of a study that aims to establish how the built environment affects the wellbeing of a building's occupants. A longitudinal framework will involve collecting an extensive set of observations, both qualitative and quantitative in nature. This would most likely start by recording field data across a spectrum of tenants in either a commercial or residential setting, including their preferences. The study would then map out basic evidence of behaviors associated with choices the tenants make around aspects of the environment they value – for example, comfort and temperature. A range of environmental quality considerations regarding building construction beyond traditional metrics of energy usage can be drawn from field data. These include behavioral measurements exemplifying thermal comfort of tenants, such as individual preferences for room temperature, which can be gauged from how often heating and cooling systems are used.

One of the limitations of longitudinal studies is their smaller scale. As pilot studies vary widely, it is difficult to draw generalized criteria of targeted impact – that is, to answer the question of when impact is "good enough." Development practitioners think of this limitation as common when carrying out resource-constrained studies. Inherently, the impact outcomes reported will be only as good as the availability of resources, such as funding, the amount of time and energy devoted to each individual pilot program and the number of beneficiaries assessed. In other words, assessing impact outcomes through such studies is constrained by the limited resources put into them. The findings may also vary depending on the timescale and how long the study runs for. In the built environment example, a tenant preference for thermal comfort is likely to vary with the different seasons or the time of day. Nevertheless, these analyses can be easily replicated, which makes them appealing to academics and industry practitioners looking to tailor the scope of larger scale projects, as illustrated in Figure 1.1. In addition, the findings of longitudinal studies can be approached with *a forward-looking orientation*. By relying on a set of studies to draw adaptive behaviors from tenants and minimize the environmental impacts they may trigger, they can be used to inform policy decisions in similar settings.

The late 1990s saw the introduction of the *Logical Framework Analysis (LFA)*, also known as LogFrame Matrix, as a tool to incorporate the context for which a specific program is evaluated. That included reviewing activities, inputs and objectives at each stage of its rollout vis-à-vis its operating environment, which enabled knowledge transfer within an organization. The LFA added the exploration of cost-and-effect relationships to an organization's impact assessment

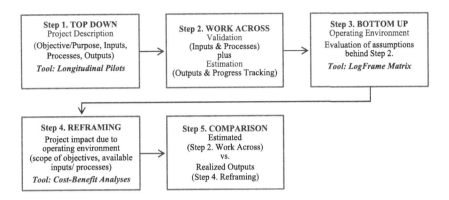

FIGURE 1.1 Five-step comparison of impact evaluation tools. Author's intuition. Reference: Barreto (2010).

toolkit, which means that potential deviations between expected and realized outcomes can be identified in a programmatic fashion.

The adoption of the LFA has opened the door for dynamic reassessment of pilot programs and made it possible to calibrate them against sudden changes in the operating environment. Its systematic analysis provides an actionable planning roadmap and builds on a common set of metrics to gauge project progress and achievements. However, the LFA's strong focus on results and the need to meet budgetary constraints may limit an organization's ability to define pivots and explore opportunities that had not emerged in the planning phase.

Stakeholder identification is an example of this – interest groups that may have not been considered at inception may not be discovered. It may also hinder innovation itself as a tool to create swift turnaround under volatile operating conditions.

THE STAKEHOLDER LENS ON METHODS AND TOOLS

LogFrame analyses brought a vital development to the evaluation of the operating context – the introduction of *participatory methods*. Such methods ensure that impact beneficiaries (the target population assessed against standalone projects and/or broad organizational commitments) are elevated from being "subjects" to the evaluation and "objects" of the assessment to become "active participants" and providers of invaluable feedback.

Participatory learning and action methods (PLA) allow diverse opinions and perspectives to be introduced over the lifetime of the assessment, ultimately weighing in on the outcome.

For example, in the field of learning and education, these analyses have been deployed to incorporate the opinions of all relevant interest groups in the planning stage of a new program. This contribution helps to better address the operating environment and its day-to-day reality. In areas of community engagement, such as poverty or access to education, adopting a PLA *mindset* helps address issues by making it a priority to learn from the active participants. Interventions can be directly shaped to the living conditions by leveraging local community knowledge – regardless of age, ethnicity or literacy.

I refer to it as PLA mindset simply because maps and other visual representations of community life enable outsiders to see the relevance of shared resources and facilities through the eyes of community members. Although part of a qualitative research toolkit based on group analysis and learning, PLA is designed to provide insights and vital feedback to validate data with all stakeholders before an action plan is made. The transect walk presented in the Technical Note provides a PLA for land use evaluation. The mapping exercise is usually conducted by walking an area with members of the local community and creating a live representation of the direct observations. Transect walks have become quite useful in collecting input from native populations to address deforestation issues and community access to transport. (See Technical Note for the set-up of a transect walk.)

It is no surprise that *stakeholder capitalism* is increasingly regarded as the natural extension of over fifty years of impact theory and responsible practices conducted outside of the private sector. I refer to stakeholder capitalism as the outright and programmatic consideration of the interests of customers, employees, suppliers and communities – in addition to those of shareholders and other financial partners – in business decision-making. When we extend impact assessment tools to a wider range of use cases by private sector participants, we are confronted with open questions about accountability: Who is ultimately responsible for delivering impact at the organizational level across all projects and initiatives? How can we go about planning for impact as a portfolio of business decisions? How do we measure, forecast, manage and adapt longitudinal studies in a forward-looking way to inform future decision-making while also benefiting from participatory environments and cross-functional cooperation? In other words, how do complex organizations manage for impact? How do they measure the effectiveness of their efforts?

Intuition of what success may look like when an accountability mechanism is put in place is usually directly related to how transparent an organization is in the ordinary course of doing business. More often than not, impact-oriented organizations make no secret of their ambitions. Transparency becomes an internal compass to help them move across functions and engage their best players. Historically, publicly listed companies have released quarterly financial statements and annual reports. This means that accountability demands on leadership and organizational structures are directly connected to meeting legal reporting obligations as opposed to defining a stakeholder approach to public disclosures.

The need to meet regulatory requirements has defined which interests are prioritized (shareholders first) instead of ensuring that an organization reports progress with a long-term lens in mind. In other words, current reporting and disclosures in the private sector incentivize boilerplate financial metrics. This comes at the expense of the organizational learning and reframing that traditional Logical Framework Analyses may suggest, and has created accountability structures that over-emphasize short-term, quantitative, financially oriented goals and hierarchical management models. Goal attainment becomes a proxy for organizational effectiveness. The reporting of impact goals and aspirations will need to shift from intermediate outcomes to learning-oriented policies. This will target improvements in organizational behaviors and metrics of organizational performance.

THE RISE OF THE IMPACT SCIENTIST

In recent years, the field of *data science for impact* has seen the influx of academic researchers as well as practitioners turned academics. Although skewed to STEM disciplines, scientists have historically excelled at unveiling less visible, hard-to-measure forms of impact through interdisciplinary studies. They have helped conceptualize the importance of seeking practical outcomes as equally valuable in a non-academic context (notably, outside of the laboratory).

Joint efforts with practitioners have facilitated the transfer of scientific knowledge into society and encouraged it to become more popular in the private sector. In turn, this has increasingly translated into more opportunities to align funding and deliver research findings that are relevant to, and accessible by, a variety of audiences. Today, impact researchers are sitting at the intersection where knowledge can be made more accessible through privately financed projects and a participatory environment can be created – one where practitioners, scientists and policymakers can interact. Yet, a handful of roadblocks still needs to be overcome.

First, the time that it takes for tools to go from lab research to commercial scalability needs to be shortened – multiyear cycles must be replaced with shorter term, action-oriented tools. Second, observations that are not readily quantifiable need to be integrated more easily. New ESG data sources rely on non-traditional metrics; for example, those addressing consumption behaviors linked to online reviews or social media coverage discussing supply chain disruptions, and customer satisfaction and brand loyalty with respect to ecolabels and other environmental claims. As with many other interdisciplinary efforts by the private sector to create actionable research findings, metrics incorporating stakeholder engagement are increasingly a sign of "high impact" research. In fact, while participatory learning remains a core foundation of applications of data science geared toward open form innovation, impact data sources and impact modeling techniques need to be both *credible* and *direct* contributors to advancing societal, economic and scientific milestones. To foster a dynamic and open dialogue between researchers and users, they must also be communicated effectively.

An example is the CONSENSUS (CONsumption, ENvironment and SUStainability) Project[1] carried out by a team of researchers from Trinity College Dublin and National University of Ireland, Galway, on behalf of the country's Environmental Protection Agency between 2009 and 2013. This was an interdisciplinary project on sustainable consumption with a call for an international team of scientists from a variety of fields to further the societal and economic challenges of advancing a consumption model for the local market in alignment with the principles of sustainable development. Most importantly, it identified a series of policy interventions and educational initiatives as transition frameworks to stir more sustainable household consumption practices in Ireland to the year 2050 and ultimately to minimize the impact on the environment.

The CONSENSUS project addressed the policy, regulatory and economic implications of varying consumption pathways across key decision areas such as energy, food, mobility and water. The purpose was to identify individual- and group-level barriers that discouraged the surveyed population from making more sustainable choices. In other words, how could 100,000 members of a local community in Ireland, along with over 100 representatives of the private and public sector, be encouraged to adopt pro-environmental actions in lifestyle-oriented practices such as heating, water usage and nutrition? The Project remains one of the leading references for how to design research efforts that promote science-policy knowledge exchange and employ foundational elements of stakeholder participation and engagement. That participation and engagement flow from the initial phases of project design to reframing the methodology of inquiry based on the initial assessment. This turns stakeholders themselves into users of the research output – in the case of CONSENSUS, as supporters of the policy actions adopted by the local government authorities. It also defined a toolkit of practical, multiyear transition pathways for consumers – whether affecting energy management, water and food consumption and waste, or transportation alternatives – that also integrates ongoing education of the local community on these topics through 2050.

THE ROAD AHEAD: DOUBLE MATERIALITY

What is ahead? This decade will create many opportunities for impact scientists to deliver actionable milestones. What the year 2020 – the year of the global outbreak of COVID-19 – showed us is that the ability to incorporate "materiality" of information in day-to-day management is a core tool for organizations to gauge both their appetite for risk and how well developed their resilience toolkit is. And this applies to organizations across sectors.

Private sector participants think of *materiality*, from its applications in the accounting and auditing professions, as information that is substantive and affects the ability of a company to deliver its business objectives. In a broader context, it is material what is *decision useful* – a characteristic that has become more frequently associated with materiality as investors have begun weaving in its assessment to identify issues outside of the traditional financial metrics and directly engage with companies on sustainability topics. This helps them to directly address those social and environmental practices that, if left unattended, may severely hinder the operational and financial future of an organization and the ecosystem(s) it participates in.

Since 2011, the Sustainability Accounting Standards Board (SASB)[2] has provided standards for accounting towards those material dimensions across sectors. Its most valuable contribution to scientists, to financial, operational and data analysts, and to entire management teams has been to articulate a mapping tool to help define material issues and associated metrics by sectors of the economy. This allows practitioners and academics to identify gaps in an organization's assessment in a transparent and accessible way. Moreover, financial market regulators globally have supported the evolution of materiality standards and their reporting by publicly listed companies by arguing in favor of the dynamic and double nature of materiality factors ("double materiality"). Let me explain.

The label *dynamic* underlines that sustainability dimensions may change over time as an organization, its sector, and the products and services it offers adapt to meet the unique needs of its end markets. Simply put, as the world around us evolves, and as production and consumption, and economic and social interactions adapt from change, so must the assessment of a company's resilience to face emerging risks and its readiness to capture opportunities.

The "double" nature of materiality is the extension of an organization's inner resilience in the context of its ecosystem. How does it react? Is it ready to weather environmental and societal uncertainties? And the opposite: In its ordinary course of business, how much does an organization help or harm the environmental, societal and economic backdrop of its ecosystem and its communities? Is it creating positive or negative externalities? For a materiality toolkit to successfully quantify those externalities, it has to be backed by data that can be translated into strategic decisions and in its day-to-day operating processes. Public and private sector organizations continue to make ambitious pledges – from reaching carbon neutrality to improving the environmental footprint of their products and operations, to achieving diversity, equity, inclusion and ethical design of artificial intelligence tools. The evolution of double materiality and the integration of data science techniques into organizational decision-making will be core to the way companies manage their impact potential, redefine their competitive advantage and build resilient organizational structures to support their double role as economic and societal agents of growth.

IMPACT AS A CRITICAL SUCCESS FACTOR

How do we reframe the way sustainability initiatives are adopted within the organizational structures of businesses to ensure continued learning beyond the narrow scope of individual project milestones? How do we achieve the ambitious goals that private and public sector participants are taking on for the new decade? What is increasingly clear, in the post-pandemic world and in light of the UN Sustainable Development Goals (SDGs), is that aligning the organizational incentives and cross-functional processes of businesses to deliver their impact-oriented commitments requires a partnership. A partnership with the scientific community will advance the work of data science for impact.

The emergence of the double materiality spectrum of sustainability risks is changing the competitive landscape and the operating environment for companies – beyond what public consciousness about these issues may suggest. Data-driven decision-making around sustainability issues is a critical success factor (CSF) for organizations. By recognizing this, organizations can stay ahead of the curve and become, inherently, more resilient. It also means adopting alternative approaches to translate scientific knowledge into action. This exercise is likely to introduce a whole new set of short-, medium- and long-term drivers that affect our collective learning, day in and day out.

One potential reason why sustainability initiatives that target impact lack the scale to succeed beyond narrow milestones may be that so far data-informed decision processes have not been consistently mapped to strategic planning. When data is removed from the strategic goal-setting exercise that most organizations run annually, it is rarely perceived as part of its growth roadmap. As a result, it is not planned for or communicated, and not allocated resources in a programmatic way – or even intentionally. What makes it even more relevant in the private sector is that remuneration packages, which are traditional organizational incentives, are hardly aligned with maximizing the full economic and societal impact potential of an enterprise. When companies' impact commitments are not connected to their organizational incentives, it becomes easy to see those commitments as aspirational instead of true levers for growth, resilience building and accountability. By recognizing the role of data as a critical success factor in the fabric of incentives and remuneration packages, it becomes more directly connected to delivering impact goals at the enterprise level. It also can potentially resolve the open questions regarding accountability.

The Critical Success Factor (CSF) method was pioneered at the MIT Center for Information Systems Research (CISR) in the 1970s. It is based on the understanding that in order to succeed, organizational goals need access to decision-useful information related to the prevailing operating environment. CSFs are defined as "the few key areas where *things must go right* for the business to flourish," and "areas of activity that should receive constant and careful attention from management" (Rockart 1979). From a pure project management perspective, that translates into assembling all information that uniquely contributes to the success of that project.

The experience of impact-oriented organizations adds to the CSF mindset the need for early incorporation of active engagement – for example, through participatory learning – in identifying which information is truly *critical*. In other words, in order for impact objectives to have a chance of succeeding, *critical* data must be identified as a representation of all critical stakeholders' contributions. Table 1.1 provides examples of CSFs, organizational goals and impact objectives in three industries.

Critical input affects the identification of impact areas for success. Therefore, it needs to be integrated with the organization's strategic objectives in management and control systems that can drive attention to all success factors equally, including impact measures. Researchers in the MIT CISR team identified four primary sources of CSFs for an organization:

1. Structure of the industry
2. Competitive strategy and geographic location
3. Regulatory pressures and geopolitics
4. Time horizon/present situation (starting point)

TABLE 1.1
Critical Success Factors, Organizational Goals and Impact Objectives

Representative Sector	Critical Success Factors (CSF)	Organizational Goals	Impact Objectives
Automotive	**Product Quality & Safety** Distribution/Dealer Network **Cost Competitiveness** Energy Standards	Market Share Product Success Earnings per Share	**Safety** Labor Practices **Design Lifecycle**
Supermarket	**Product Mix** **Inventory Management** **Price Competitiveness**	Market Share Product Success Earnings per Share	**Customer Satisfaction** Labor Practices Community Recognition
Hospitals/ Healthcare Delivery	**Facilities Management** Resource Efficiency **Cost of Healthcare Delivery**	Excellence of Care Patient Population Optimal Procurement	**Access & Affordability** **Health & Safety** Personal Data Protection

Source: Author's own research and institution.
Bold text denotes direct mapping between impact goal and critical success factors of companies operating in the sector.

Sources #2 and #3 highlight how organizations operating in the same ecosystem and similar markets may exhibit distinct CSFs. These may also vary, depending on the unique starting point in which organizational goals are set and, therefore, where and when impact assessments start taking place (source #4).

Historically, the identification of CSFs has relied on interviews with a company's leadership team and, occasionally, their key personnel. But these interviews have ranged widely from an open-ended exercise to more a meticulously structured collection of information. An evaluation focused on a CSF mindset is distinguished by its ability to go beyond the easy-to-collect data. It can focus on the information that otherwise may not have been collected, either because it was not historically part of the existing management and planning system used to make decisions, or possibly because it was overlooked.

CSFs are dynamic in nature and organizations operate under changing conditions. Together, these facts highlight the relevance of taking a holistic approach to building the theory behind CSFs to incorporate primary impact areas – those that are most likely to yield both stakeholder interest (internal and external) and leadership support as a path to strengthen organizational resilience and achieve stated impact goals.

Analytical methods are becoming a must for organizations seeking to find their unique path to incorporate impact factors in ordinary decision making and make this an integral part of their success story. An important first step is reasoning in terms of CSFs to embed impact in organizational design. This allows outcomes to be clearly visualized and creates a unique compass. A starting point on the roadmap can be a product, a process or simply a customer request that helps better calibrate "success" and supports the authentic integration of impact-oriented data.

Defining a sound structure that enables continued learning while driving outcomes is a lasting proposition and a worthy quest of the impact enterprise. Not to mention the fact that, without a uniquely tailored organizational structure to drive incentives, accountability and performance management, private or public sector participants may not be willing to continue paying for services and products that lack societal and economic relevance. Models of redistributed accountability for business outcomes usually suggest that when a factor is most critical for a company's success, it is not necessarily best confined to a single department or function. This is because such factors permeate all decisions and behaviors across the organization.

The media rarely points to role models in innovation that have thousands of patents but no research and development department. Only a handful of the most widely recognized employers for whom talent acquisition and retention are core do not have a dedicated Learning and Development team. This may be because these businesses have aligned CSFs with a core mission statement, which integrates their target impact across all organizational layers and makes each decision critical to deliver business outcomes. But they are the exception. Many organizations continue to diversify their offerings, their geographic locations and the customers they serve with the single goal of increasing the stability of their financial results. In those instances, adopting analytical tools without organizational support is likely to yield inconclusive outcomes and, therefore, virtually no learning potential.

A NOTE ON ORGANIZATIONAL STRUCTURES FOR IMPACT

Organizations that make early commitments to deliver "impact outcomes" often establish internal task forces that bring together multidisciplinary knowledge and cross-functional expertise to aid with strategic direction. However, while internal efforts enable early buy-in, they may also contribute to diluted accountability. This makes it hard to assess how effective an incentive is and potentially slows down decision-making. But by recognizing that sustainability dimensions can be critical success factors for the business, the integration of impact objectives is best promoted by designing information sharing systems, introducing specialized coordination roles and deploying standardized routines. This means that participatory learning and changing conditions in a company's operating environment can be incorporated through open visualization platforms.

The experience of international NGOs and the public sector in the evaluation and assessment of project-level impact offers a series of time-tested approaches that can directly benefit sustainability efforts of businesses. The increasing availability of data science tools and the wider acceptance of multidisciplinary frameworks for impact represent the foundation for companies to reframe operational processes that integrate sustainability dimensions as critical success factors. So where to start when designing the organizational ecosystem that promotes sustainability objectives within a business and interweaves talent, data, systems and accountability to deliver impact goals? By tackling the questions:

How does the organization learn?

What triggers long-lasting learning?

THE LEARNING JOURNEY – THE IMPACT ENTERPRISE

The application of analytical methods and participatory learning techniques can address business challenges and reshape the way organizations learn and act on their impact objectives. The following list provides a step-by-step roadmap to start identifying the sustainability needs of a business of your choice, and develop a suitable lens for double materiality, by deploying the impact evaluation tools introduced earlier.

- Compile a list of the sustainability commitments of the business, including both societal and financial targets and the timeline of these commitments (publicly disclosed or from internal discussions). Identify potential impact assessment tools (e.g., cost–benefit analyses, longitudinal studies, etc.) that may be already in use within the organization. Address pros and cons associated with each and the areas where the tools may be delivering a backward-looking evaluation of progress, not a forward-looking assessment. Take note of the gaps.
- Research the operating context of the business by thinking of the critical success factors that underpin the sector overall. Which factors may evolve faster than the proposed time horizon for the delivery of the impact commitments?
- Design a transect walk. Identify the stakeholders (internal or external) actively involved in the realization of impact. Employ active listening to survey the participants in redefining the impact objective and business commitments through their lens.
- Based on the input from the transect walk, highlight a set of critical success factors for the business that best describe its unique operating environment and prioritize impact commitments on a timeline. Take note of traditional accountability structures by function versus the organizational incentives or coordination mechanisms that would favor shared knowledge and faster delivery of impact outcomes. Any lesson learned?

TECHNICAL NOTE – THE DESIGN OF A TRANSECT WALK

UK-based conservation charity Fauna & Flora International (FFI) defines a transect walk as "a tool for describing and showing the location and distribution of resources, features, landscape, and main land uses along a given transect." FFI encourages the use of a transect walk to better articulate any causality among "topography, soils, natural vegetation, cultivation, and other production activities and human settlement patterns" (Fauna & Flora International 2013, 1).

As a practice, it allows organizations to gather spatial data and identify any roadblocks along a dedicated area under evaluation (for example, that of a transect along a project site), including social and environmental vulnerabilities or outright threats. It stirs dialogue about how communities along the transect may seek to resolve an issue (whether in terms of ecosystem protection or by pointing to access to key natural resources or transportation sites for communities).

As a pure mapping exercise, a transect walk is usually designed as a collaborative form of data gathering undertaken along a predefined geographic site by a researcher with representatives of the local community. While walking, the group sketches the surrounding areas – conditions of transport, state of the territory, access to waterways, distribution and use of shared lands, and so on. This is done in a way that reflects any hardships involved in carrying out daily activities or the recounting of a particular event from the perspective of the local inhabitants. Usually, the selection of the representatives will reflect if there are any biases towards a particular segment of the local population.

The FFI suggests an average of three hours should be devoted to engaging in the transect walk. But the duration of the exercise will depend on the quality of the information gathered as well as the visual observations captured while en route, such as estimation of distances, weather patterns, elevation and state of existing infrastructure.

The example presented in Table 1.2 shows a transect walk as part of a broader biodiversity assessment. First, a sketch of the surrounding areas aims at mapping key elements of the natural environment (in this case, the presence of wildlife, lakes and rivers, agricultural lands and living quarters, and any transportation hubs). Second, leveraging the dialogue with the local representatives, a description of the benefits to the community stemming from the existing layout (e.g., well-defined land use and separation of use between dedicated areas), and its potential limitations (e.g., access to essential facilities and transport), allows the researcher to start identifying present and/or future hurdles from direct observation and the recounting of the community along the walk.

As the mapping is quite collaborative in nature, discussing alternatives and solutions may be a valuable form of engagement with community leaders as well, or may be brought about for inquiry at a later time. It is important to note that usually up to three walks are needed to fully capture the area under review, potentially exploring different routes, and to engage with different stakeholders along the way as their value is highly dependent on the ability of the researcher to adapt the walk to the local context.

TABLE 1.2
Example of Transect Walk

	⚲ Nature	⊞ Built Environment	🚃 Human Activities
Land use	Description and inventory of wildlife, water resources, trees and shrubs.	Position and state of villages, houses, construction sites and other buildings.	List of economic and cultural activities (agricultural, commercial, residential, recreational).
Access to facilities and transport	Description and inventory of transportation hubs (land/water) and degree of accessibility.	Position and state of utilities for residential and agricultural use (infrastructure and services).	List of access enablers and/or lack thereof. Closeness to transport, storage and other common facilities.
Walk observations	Note signs of deforestation, water pollution, condition of forests and aquifers.	Record waste water and irrigation mechanisms, sewage/domestic dwellings.	Note conditions of access to and from facilities, to and from built environment vs. nature and land resources.

NOTES

1 https://www.tcd.ie/Geography/research/environmental-governance/projects/consensus/.
2 Known as the Value Reporting Foundation since the merger of SASB with the International Integrated Reporting Council in June 2021.

REFERENCES

Benzer, J. K., M. P. Charns, S. Hamdan, and M. Afable. 2017. "The Role of Organizational Structure in Readiness for Change: A Conceptual Integration." *Health Services Management Research* 30 (1) (February): 34–46. doi:10.1177/0951484816682396.

Bond, A., and C. Chanchitpricha. 2013. "Conceptualising the Effectiveness of Impact Assessment Processes." *Environmental Impact Assessment Review* 43 (November): 65–72. doi:10.1016/j.eiar.2013.05.006.

Bullen, C. V., and J. F. Rockart. June 1981. *A Primer on Critical Success Factors.* Massachusetts Institute of Technology, Center for Information Systems Research, CISR no. 69, Sloan WP no. 1220–81. https://dspace.mit.edu/bitstream/handle/1721.1/1988/SWP-1220-08368993-CISR-069.pdf?sequen.

Calace, D. 2020. "Double and Dynamic: Understanding the Changing Perspectives on Materiality." *SASB Blog*, September 2, 2020. https://www.sasb.org/blog/double-and-dynamic-understanding-the-changing-perspectives-on-materiality/.

Cuypers, I. R. P., P. Koh, and H. Wang. 2016. "Sincerity in Corporate Philanthropy, Stakeholder Perceptions and Firm Value." *Organization Science* 27 (1): 1–231. doi:10.1287/orsc.2015.1030.

Douglas, E. M., S. A. Wheeler, D. J. Smith, I. C. Overton, S. A. Gray, T. M. Doody, and N. D. Crossman. 2016. "Using Mental-Modelling to Explore How Irrigators in the Murray–Darling Basin Make Water-Use Decisions." *Journal of Hydrology: Regional Studies* 6: 1–12.

Ebrahim, A. 2005. "Accountability Myopia: Losing Sight of Organizational Learning." *Nonprofit and Voluntary Sector Quarterly* 34 (1): 56–87. doi:10.1177/089976400 4269430.

Fauna & Flora International. 2013. "Transect Walk: Conservation, Livelihoods and Governance Programme Tools to Participatory Approaches." February 2013. https://earthrights.org/wp-content/uploads/transect-walk.pdf.

Hynes, M., A. Davies, F. Fahy, H. Rau, L. Devaney, R. Doyle, B. Heisserer, M. Lavelle, and J. Pape. 2015. "ConsEnSus: Consumption, Environment and Sustainability." doi:10.13140/RG.2.1.3499.5043.

Jasny, B. R. 2013. "Realities of Data Sharing Using the Genome Wars as Case Study – An Historical Perspective and Commentary." *EPJ Data Science* 2 (1). doi:10.1140/epjds13.

Kourany, J. A., and P. M. Fernández. 2018. "A Role for Science in Public Policy? The Obstacles, Illustrated by the Case of Breast Cancer Screening Policy." *Science, Technology, & Human Values* 43: 917–943.

Langevin, J. 2019. "Longitudinal Dataset of Human-Building Interactions in U.S. Offices." *Scientific Data* 6: 288. doi:10.1038/s41597-019-0273-5.

Lydenberg, S., J. Rogers, and D. Wood. 2010. "From Transparency to Performance." Harvard University Initiative for Responsible Investment.

Macilwain, C. 2000. "Biologists Challenge Sequencers on Parasite Genome Publication." *Nature* 405: 601–602. doi:10.1038/35015216.

Middlemiss, N. 2003. "Authentic Not Cosmetic: CSR as Brand Enhancement." *Journal of Brand Management* 10: 353–361. doi:10.1057/palgrave.bm.2540130.

Napier, A. and N. Simister. 2017. "Participatory Learning and Action." INTRAC, https://www.intrac.org/wpcms/wp-content/uploads/2017/01/Participatory-learning-and-action.pdf.

Nordhoff, S., J. de Winter, R. Madigan, N. Merat, B. van Arem, and R. Happee. 2018. "User Acceptance of Automated Shuttles in Berlin-Schöneberg: A Questionnaire Study." *Transportation Research Part F: Traffic Psychology and Behaviour* 58: 843–854. doi:10.1016/j.trf.2018.06.024.

Oreskes, N. 2004. "Science and Public Policy: What's Proof Got to Do with It?" *Environmental Science & Policy* 7 (5): 369–383. doi:10.1016/j.envsci.2004.06.002.

Porter, M. E., and M. R. Kramer. 2002. "The Competitive Advantage of Corporate Philanthropy." *Harvard Business Review* 80 (12): 56–69.

Rau, H., G. Goggins, and F. Fahy. 2017. "From Invisibility to Impact: Recognising the Scientific and Societal Relevance of Interdisciplinary Sustainability Research." *Research Policy* 47 (1): 266–276. doi:10.1016/j.respol.2017.11.005.

Roche, C. J. R. 1999. *Impact Assessment for Development Agencies: Learning to Value Change*. Australia: Oxfam.

Rockart, J. F. 1979. "Chief Executives Define Their Own Data Needs." *Harvard Business Review* (March). https://hbr.org/1979/03/chief-executives-define-their-own-data-needs.

"Sacrifice for the Greater Good?" 2003. *Nature* 421: 875. doi:10.1038/421875a. https://www.nature.com/articles/421875a#article-info

Sánchez-Andrade Nuño, B. 2019. *Impact Science: The Science of Getting to Radical Social and Environmental Breakthroughs*. USA: Creative Commons.

Senge, P. M. 1990. *The Fifth Discipline: The Art and Practice of the Learning Organization.* New York: Doubleday/Currency.

Sustainability Accounting Standards Board, SASB Materiality Map®, https://www.sasb.org/standards-overview/materiality-map/.

The World Bank. 2000. *The Logframe Handbook: A Logical Framework Approach to Project Cycle Management.* Washington, DC: The World Bank.

2 Organizational Learning for Impact[1]

Chapter 1 defines *organizational learning for impact* as the programmatic learning established by companies seeking to define their impact objectives and deliver business and societal value using data analyses and analytical frameworks of participatory learning. Organizations follow a handful of learning models, which are highly correlated to their organizational structure. An organization's culture and what it perceives as evidence of success also shape the way it learns. To be effective, learning must be underpinned by a data-oriented culture and an organizational structure that thrives on analytical investigation and subsequent inquiry of outcomes. This means that the first step to creating a lasting roadmap for learning by doing – and ultimately, for creating enterprise value – is to establish enterprise-level data methods to address environmental and social challenges and unveil the leverage points of a business.

Where shall we look to design such an impactful and essential system? One that comprises talent, processes, data, systems and accountability toward sustainability objectives. By starting with the relationships that form the inner fabric of decision-making. By answering the question "How does the organization learn?" or "What is the learning culture of the system?" we can articulate what triggers long-lasting learning.

I often think of organizations as falling in one of the following categories:

A. innovation-oriented
B. manufacturing and process improvement-oriented or
C. a combination of A+B

The vast majority of businesses are probably a combination of the first two categories. This could be because of their existing context or their historical upbringing. Or it could be because they aspire to innovate their business model to adapt or become more agile and effective in the way business outcomes are delivered, how they are perceived by their consumers and how they meet that population's evolving needs. This is particularly true in the private sector. As businesses add impact-oriented goals to their external and internal commitments, they inherit a modus operandi that reflects the way they traditionally do business. That is the premise and the context in which societal goals may simply be added to the anonymous bucket of business deliverables. Should those goals be addressed the same way as other deliverables, or should their organizational context be reframed? Let us explore this further by looking at a company that has paved the way toward long-term value creation through sustainability.

DOI: 10.1201/9781003212225-3

AN EMPIRICAL STUDY OF LONG-TERM VALUE CREATION

Japanese R&D-based global pharmaceutical giant Eisai Co., Ltd., is a pioneer in the *human health care (hhc)* space. It offers a distinct example of both the journey that a company needs to take and the organizational perspective it needs to address when adding the impact dimension to its goals.

Dr. Ryohei Yanagi, Ph.D., is a Visiting Professor at Waseda University, Graduate School of Accountancy, in Tokyo and Chief Financial Officer of Eisai Co. Ltd. The company is a leading global manufacturer and distributor of prescription drugs. Its mission, as expressed in its Articles of Incorporation and shown in Figure 2.1, is literally to humanize healthcare for patients and their families. Under his expert leadership, Dr. Yanagi has turned the company's global footprint into an example to follow when connecting impact assessment methodologies to financial performance in a publicly listed company.

The data methodology developed by Dr. Yanagi is instrumental in driving Eisai's commitment to hhc and bringing along a deeper organizational learning. Instead of focusing on the potential data availability challenges, which could prevent the learning process itself, Dr. Yanagi built a framework to bring influence to Eisai's corporate mission. He unveiled analytically the hidden intangible value that companies such as Eisai (and their financial stakeholders) would be able to tap into by pursuing a strategic orientation toward material environmental and social governance (ESG) factors in its global businesses. (See Technical Note for a summary of the Yanagi Model.)

As an expert in Japanese corporate governance, Dr. Yanagi's findings have been featured in prominent forums worldwide. They continue to provide an example of what is possible when organizational learning around non-financial, material metrics is incorporated into the decision-making of a company's operating culture – a culture committed to the pursuit of social and environmental performance along with operating profit. Specifically, Dr. Yanagi helped financial and business practitioners

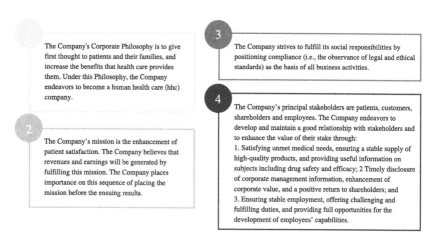

FIGURE 2.1   Articles of incorporation: Eisai Co., Ltd. (Eisai 2020, 14).

realize how investments in human capital (traditionally labeled as "labor costs") and intellectual capital ("R&D expenses") contribute to enhancing corporate value and, consequently, financial returns over the medium to longer term.

I asked Dr. Yanagi to share an example of the value of his empirical research, in particular, how his methodology captured societal and financial goals in the context of his organization's hhc mission. In his answer, he quoted the organizational learning roadmap as a key differentiator. Introducing key performance indicators (KPIs) of social and human capital (such as labor costs, female representation in the workplace, promotion rate of female employees and employment rate of workers with disabilities) strongly reinforced the dialogue with the firm's unionized labor while directly linking workers' wellbeing to the ability of the business to deliver positive financial returns to its investors. In Dr. Yanagi's words, a "win–win" situation where intellectual and human capital directly add to the potential value to shareholders.

Whether in the context of impact goals or not, it is no surprise that learning processes are dynamic. According to MIT Professor Emeritus Ed Schein, there are different types of learning, and they have different time horizons. This inherently allows learning to evolve as the organizational context and the operating conditions of the enterprise change. A knowledge acquisition step adds to our knowledge base. We develop the ability to gather insights, which become useful as our knowledge base deepens and we see how it can apply across a broad spectrum of potential situations.

Regardless of whether "impact" thinking has been already woven into the fabric of an organization, such as in the case of Eisai's corporate mission, it requires leaders to frame a new vision. They must address those "temporal and spatial invisibilities" that create inevitable delays when trying to establish both causality and correlation between impact-oriented decisions and their measurable outcomes. In the case of Eisai, Dr. Yanagi refers to these structural delays between strategic decisions and organizational value in terms of the number of years it takes for specific KPIs of non-financial capital to correlate with measures of financial performance. According to Prof. Schein:

> When leaders articulate a new vision for their organization and communicate that vision widely, they are typically trying to give large numbers of people in the organization a new insight, and, if they are successful at this, the organization can change directions rather quickly. Developing a new vision and sharing that vision widely can be thought of as one necessary step in speeding up learning.
>
> *(Schein 1992, 3)*

How does complexity work in environmental and social systems? Systems that have lots of connections and interactions, involving trade-offs, feedbacks and surprises that mean we do not see the results when we expect to. In a complex system, it is hard to understand the implications for all capital assets (both financial and non-financial, such as human capital, natural capital, and so on). Complex systems can generate both positive and negative feedback that can strengthen or hinder our sustainability efforts. There are temporal and spatial invisibilities – events that have consequences that the initial decision maker or agent may not be aware of. They show up somewhere else on the map. There is a need to map connections across scales and times and design solutions that take this into consideration.

Adaptation requires a lot of humility along with innovation and skill. In his model, Dr. Yanagi suggests taking advantage of structural delays between the integration of ESG KPIs into business strategy and the realized long-term value for a company. He suggests that one way to do this is to use the delays to deepen the learning potential by communicating the KPIs more broadly and seeking input from a wide range of stakeholders, including financial partners. Dr. Yanagi's model embeds learning as a social process by tracking progress against organizational KPIs. In summary, the delayed learning that an organization may experience presents an opportunity to better define an impact methodology to assess both financial and non-financial outcomes. The reason why the Yanagi Model is so powerful is that it provides tangible drivers of impact goals while also powering up the learning potential of a complex organization by addressing the inevitable trade-offs of long-term versus short-term decision-making. In fact, "impact goals" are often thought of as quintessential long-term goals.

Over the years, in my interactions with Dr. Yanagi, I have asked myself whether businesses first need to do some *unlearning* before learning how to embed impact as an organizational goal. Academic and practitioners' work on organizational learning abounds, but this is not the case when it comes to organizational unlearning – how to conceptualize it and put that meaning to work.

HOW MENTAL MODELS AFFECT IMPACT OUTCOMES

Mental models have often been described as internal representations of concepts and ideas, as memory constructs of our mind that enable a deeper understanding of new information. In this context, I refer to the definition of mental models by Prof. Peter Senge as: "deeply ingrained assumptions, generalizations, or even pictures and images that influence how we understand the world and how we take action" (Senge 1990, 8).

Mental models drive the interpretation we give to a concept or an idea: the way we frame a problem. Mental models can help practitioners build intentionality. They help us to visualize impact-oriented outcomes that may be more difficult to grasp through logical reasoning because of all the "spatial and temporal invisibilities" discussed earlier. Impact is an unstructured problem. What I mean by that is that societal challenges often present themselves in a multidisciplinary form, mostly at the intersection of public policy, business decision-making and science. The result is ambiguous debates among domain stakeholders that bring in quite different lenses, especially when recognizing their distinct frames of reference and the different mental models of each field. This exposes difficulties in addressing each other's interests and sharing perspectives.

There needs to be a strong effort to bridge the gap between scientific information and the insights provided by nonacademic practitioners in the context of policy

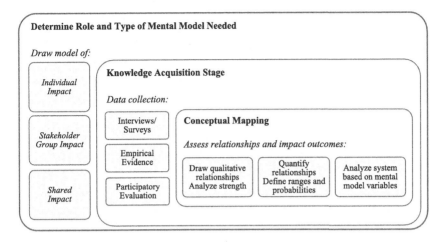

FIGURE 2.2   Mental models for impact-oriented outcomes. The author referenced the following study: Moon (2019).

or strategic business decisions. Mental models can play a key role in reaching a consensus on a mutually acceptable roadmap. Figure 2.2 offers a visual representation of the pivotal role that mental models can play in delivering outcomes. By clearly identifying the thoughts and ideas that emanate from targeting a selection of impact objectives from the perspective of individual participants and of stakeholder groups, a unique area of overlapping interest emerges (a model of shared impact). From that area, a data collection phase can better drive knowledge acquisition at the organizational level that aims at a conceptual mapping of existing relationships and tools at hand to describe their contribution to meeting a set of shared impact objectives. Viewed through that lens, mental models can become operational frameworks that incentivize the sharing of beliefs, assumptions and data points and encourage discussion of all potential alternatives in a cooperative way.

Introducing a mental model for impact is a key step in drawing an effective and transparent roadmap for learning and knowledge sharing within an organization. This is especially true when, historically, business goals have favored economic variables and short-term targets at the expense of the consideration of long-term environmental and social dimensions.

The definition of an impact outcome varies among stakeholders and is affected by whether they are directly responsible for executing an impact-oriented strategy at the organizational level or are simply recipients of that process. More often than not, these two groups overlap only to a minor degree. A lot of energy and resources can be spent on establishing a dialogue between stakeholder groups that diverge in the attitudes, norms and cultures that surround a set of outcomes. In particular, because of people, time or budgetary constraints, many stakeholder groups become polarized when faced with a resource allocation problem that affects the environment, the economy and society. The issue usually emerges between the decision makers and the recipients of the allocation.

Consider, for example, the distribution of forest lands, deforestation activities and protection of forests, in contrast to the issuance of mining permits, cycles of infrastructure and urban versus agricultural expansion. Populations affected by depletion of natural resources will view things differently from local government authorities. The construct of the mental model is also affected by the misalignment of short-term versus long-term effects of allocations that are skewed to one individual or group.

REFRAMING RESOURCE ALLOCATION
PROBLEMS THROUGH MENTAL MODELS

Let us now look at an example and see how mental model mapping can be applied to allocating water for irrigation purposes. In regions where agriculture is a major component of socio-economic wellbeing, there are local regulations governing natural water sources and their use. Farmers also play a large role in decision-making, regardless of whether that decision applies to planning for annual crops or more extended use, and ultimately affect the sustainability profile of the area. Yet water decision-making is quite uncertain and lacks transparency.

In the last decade, the need to facilitate knowledge sharing to identify common goals has become more relevant. The knowledge to be shared encompasses an entire region's environmental sustainability and individual farmers' socio-economic decisions. In fact, as reported by the World Resource Institute, in early 2015 the MIT Integrated Global System Model Water Resource System (IGSM-WRS) estimated that water stress is expected to affect over half the world's population by 2050 (Schlosser 2014). A mental modeling approach carried out through techniques such as fuzzy cognitive mapping and software applications such as the one offered by Mental Modeler[2] is likely to address information gaps among stakeholders. This should also encourage group learning on the socio-economic challenges that the global issue of water scarcity creates in a localized setting.

One pressing example is that of the Murray–Darling Basin region in Australia. Water sustainability has long been a concern there and has been instilled in a series of country-level policy decisions on water usage. These decisions have been steered to establish a transparent process to govern access to water and incentivize users to adopt water efficiency mechanisms. Douglas et al. (2016) employed mental modeling techniques to help coordinate action and promote a baseline for understanding the socio-economic stress that ensuring sustainable water resources creates locally. From their analyses it emerges that when adopting the lens of the farmers on water decisions and adding environmental factors to basic irrigation needs of the farmland, policymaking can become a more direct and intentional exercise built on dialogue and cooperation.

In building the mental model of water decisions, nonquantitative variables such as the personal experience of the farmers in defining seasonal forecasts are likely to play a pivotal role in the effectiveness of public sector intervention. In fact, the mental mapping done by Douglas et al. (2016) revealed how the "desire to farm"

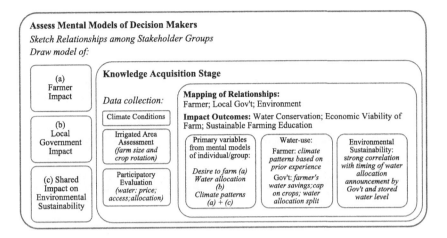

FIGURE 2.3 A mental model for water-use decisions. Author referenced the following study: Douglas (2016).

greatly influenced decision-making. Assuming continued positive catalysts for the farmer (such as age, risk appetite and farm size) and transparent near-term policy impact on farmers, it was discovered that the farmer would choose to continue farming even when faced with poor environmental conditions such as low rainfall or low water storage levels. The farmers would choose to farm over not farming even if they were compensated for water savings by the local municipality. In fact, conditions on the structure of incentives (such as water price level and variability, monetary incentives associated with water saving behaviors, etc.) play an important role when water allocation is perceived as a given and subject to low to moderate fluctuations.

The goal of applying cognitive mapping to visualize the influence variables on decision-making is to identify the relationships between variables that may affect the perspective of the decision maker – in this case, the farmer. The next step is to compare these variables with the impact of a set of policies on farmers' decisions (and mental model) to advance optimal solutions for both the near and long term (Figure 2.3).

Mapping a mental model of stakeholder interactions in the allocation of resources is the first step in learning effectively about a problem and building a foundation for shared learning.

LEARNING STRATEGIES FOR IMPACT

Innovation-oriented companies (A) offer examples of organizational "unlearning" and the positive relationship that exists with innovation. Whether impact outcomes are expressed as environmental or societal goals, they are increasingly connected with how fast a sector or a specific company is bringing innovation forward. Many of those outcomes are either delivered through technological innovation or entail a degree of new process or new product development. High-tech organizations continue to make headlines for their adherence (or lack thereof) to governance practices. They are increasingly operating at the intersection where scale-oriented business models collide with certain human sensitivities. Such sensitivities include the deployment of intelligent machines, the use of artificial intelligence, the future of work, cyber risks, the violation of privacy through the sharing of personal information and basic human rights such as freedom of speech.

This is the case in the software and computer services industry. Since early 2000, the United Nations, through its Global Compact initiative, has set forth ten principles for the implementation of responsible business policies, which include corporate complacency. But allegations of noncompliance with international human rights principles continue to damage the industry's reputation. Incidents relate to poor employment conditions, employment discrimination and gender inequality, racial inequity, union relations and the bargaining power of organized labor.[3]

High-technology enterprises have thrived by introducing a stream of process and product innovations, which inherently affect the way individuals, organizations and communities carry out sense-making. How they communicate and perform basic to more advanced tasks is replaced with faster, easily scalable and more economical alternatives. Yet, the success of an organization is inherently built on internal "facilitator" or catalyst roles, which support the launch and spread of innovation within and outside traditional organizational boundaries by leveraging social networks.

This is increasingly relevant as "radical" innovations are introduced that create entire new subindustries. This makes others obsolete and disrupts the existing technological trajectory of the ecosystem, thereby creating strong social influence. As researchers of radical innovation from the National Dong Hwa University in Taiwan point out, in the case of high-technology industries, *disrupt* means *unlearn* in the sense that a firm must unlearn its past knowledge or skills and learn new knowledge or skills that did not previously exist within its organizational boundaries" (Yang 2014, 152). It is my opinion that we are geared for unlearning as our society is ripe for disruption across a variety of fields and behaviors, many brought about by the needs of a global sustainability transition.

Unlearning comprises two dimensions: discarding (something) and replacing (something) with (something new). For the most part, changes in routines seem to fall under the "replacing" category. Yet aligning societal goals and outcomes and elevating them to the same height as traditional financial dimensions requires organizational unlearning by "discarding." That is, by replacing or upgrading processes, products and services that harm non-financial dimensions – or to put it another way, whose absence directly creates positive societal synergies (e.g., labor relations and data-driven negotiations in the case of Eisai).

An important element of the process of unlearning by discarding is addressed well by taking the perspective of resource allocation coupled with a strong social influence brought about by internal facilitators. I will call these "impact catalysts." Unlearning is a big component of "learning" in impact-oriented organizations. Impact activities, routines and beliefs are as influential nowadays as many of the disruptive technologies that help impact to materialize. In that context, we should advocate a structure that relies on a social influence network of "impact catalysts" as agents of impact both within an organization and outside its boundaries. We should also introduce an "outside in" perspective. How do we define the contribution of "unlearning" on the value created by the enterprise?

One question that comes to mind when addressing unlearning in organizations such as the ones highlighted above is their size. While most academic studies control for firm size effects (which also correlate with resource availability) when presenting scientific evidence, this is less important when it comes to impact-oriented organizational learning. In fact, in this case, it is more about whether innovation and corporate development functions that promote partnerships – also usually the ones tasked with prototyping and commercializing ideas with large societal impact – deploy a learning strategy to deliver impact outcomes.

A NOTE ON RESPONSIBLE INNOVATION

Microscale innovation accelerators, which frequently adopt unlearning as a key step in value creation for their participants, are increasingly considered as catalysts of organizational learning. Are these socio-technical knowledge networks more or less open than institutional functions insulated within a corporate environment?

The intuition presented by researchers at Delft University of Technology in the Netherlands suggests that "responsible innovation" is a leading driver of openness in knowledge networks in the context of early-stage spin-off entities from academic hubs when the focus of the innovation being brought to market is to solve sustainability challenges. While the main characteristic of companies participating in the study is their limited resources (finance, project management and time constraints), the upside of such blended ecosystems is the ability to break the path-dependency of learning (the "way to do things") experienced in many of their larger, established peers.

Not surprisingly, sustainability offers tremendous opportunities for entrepreneurs to create value in areas such as health, food/consumption, energy/water and the environment. The world of "responsible entrepreneurship" has already emerged. This is where the process of innovation addresses societal needs, and the marketability of products and services is directly linked to the positive ways they contribute to the general public. It has found renewed support in a variety of geographies, from the European Innovation Strategy for 2020 to China's Science and Technology Roadmap to 2050. The learning dynamics that characterize an innovation cluster – within and outside an academic ecosystem – rely on a highly interactive process and involve many stakeholders. The structure of embedded incentives in the organizational profile of these clusters facilitates knowledge sharing and transfer among all cluster participants. This enables significant open networks to take hold and open innovation to spread wide.

Nevertheless, it takes time for a knowledge network to evolve and to build a diverse cohort of participants. Academic accelerators usually involve university researchers and scientists in the early stage of knowledge network development. Partners more likely to add value in the commercialization/go-to-market phase usually come in during a second iteration of the knowledge network. Openness to new partners is key to accumulating transfer knowledge capacity and bringing in timely metrics of impact that may not have been introduced during the early days of the innovation cluster. These provide complementary assets. An example is the funding and commercialization of green hydrogen solutions at scale. The technology development cycle from lab-scale to industrial-scale is inevitably linked with delays in the launch of second-generation services.[4] Ultimately, innovating with a responsible, sustainability-oriented mindset requires the ability to benefit from an open knowledge network. Yet setting a vision is not enough. Just as our cognitive capacity as individuals is limited, so is that of an organization in aggregate.

That is also one of the reasons that *systems thinking* – the ability to identify, visualize and find value in highly dynamic settings – has become a widely spread toolkit. Analytical tools must allow us to go past individual limitations and break down linkages and connections underlying day-to-day, real-life events and layered organizational decision-making. When businesses start on their sustainability journey, systems thinking provides a vital first step in mapping mental models and incorporating the complex new set of environmental and social dimensions that affect an organization beyond financial and economic relationships. Thinking in systems becomes a knowledge acquisition step of its own. One of the strongest intuitions behind it is Prof. Schein's view about the adaptive learning journey of leaders:

> Such learning can be speeded up if leaders become more marginal in their own organizations and spend more time outside their own organizations. They cannot obtain insights into the limitations of their organizational cultures unless they expose themselves to other cultures – national, occupational, and organizational.

> *(Schein 1992, 14)*

Deploying analytical methods and organizational experiments, such as the one highlighted by the Yanagi Model, plays an essential role in breaking down the virtual cycle of potential organizational anxiety towards *change*.

ADAPTIVE LEARNING FOR SUSTAINABLE ENERGY INNOVATION

In my multidecade career as an investor, I have witnessed how, almost by construction, demonstration projects that involve new energy technologies are often organized as closed ecosystems of project developers, research and private sector organizations. These installations go through cycles of testing, monitoring, evaluating and improving intermediate outcomes for the purpose of scaling innovations commercially. They are designed as participatory settings where learning is shared, results are transparent and feedback is incorporated in a timely manner.

Recent studies have found that organizational learning may vary by project phase.

- Initially, participants learn to capture intellectual property and build prototypes.
- Later, they learn to build production plants and exploit learning curves.
- Finally, participants learn to create demand-supply networks.

Sustainable energy policy builds on the key lessons learned in demonstration projects that experiment with emerging technologies. Specifically, learning is ignited and maintained by the processes of interaction between project participants. This is because these interactions provide the opportunity to develop new applications or refine original project plans. It is also a function of shared learning from early failures.

What makes these types of projects quite interesting from an organizational learning perspective is the underlying socio-technical knowledge. Participants look for potential commercialization of the initial idea and rely on trust and value networks between researchers, engineers, field experts, community leaders and other public sector officials.

One application of particular interest is the adoption of low carbon technology demonstration projects, such as those following the advances in carbon capture and storage technologies. Research (Heiskanen et al. 2017) has found that, as a result of the lack of robust *prior learning*, participants in such early-stage demonstrations may struggle to implement scientific knowledge that is also quite novel. This may be because the field is a niche market and there are limited early adoption opportunities that can be scaled further. However, the delay between the first demonstration project and the next (the temporal invisibilities I introduced earlier), makes it harder to identify the most effective learning strategy for project participants. In turn, this makes it harder to transfer knowledge in the following stages of project development.

Learning is often quoted by practitioners as the leading motivation for stakeholders to join early-stage demonstrations. This makes the shared knowledge transfer just as beneficial to all participants as the future production of the technology itself. In fact, it allows technical feasibility issues to be solved and creates the opportunity to experiment with new markets through a sub-series of demonstration projects. Therefore, by investing resources in these novel efforts, organizations are more likely to apply the learning to the deployment of future technologies. Fieldwork led by Bart Bossink in 2020 attributes this learning

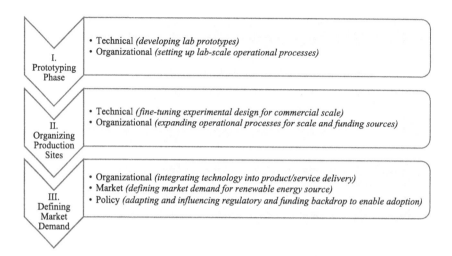

FIGURE 2.4 Learning strategies for renewable energy demonstration pilots. Author referenced the following study: Bossink (2020).

cycle and the set of facilitating behaviors in renewable energy pilot projects to four learning strategies. These strategies follow the evolution of the prototyping/organizing/marketing phase of each project (Figure 2.4):

- *Technical* learning (e.g., development of scientific/engineering knowledge of the underlying technology)
- *Organizational* learning (e.g., development of new production systems and logistics/delivery of the underlying technology)
- *Market* learning (e.g., development of knowledge of the target customer needs)
- *Policy* learning (e.g., development of regulatory and public policy knowledge of the underlying technology and its deployment/roadblocks-opportunities)

The example of evolving learning milestones in the creation of value for renewable energy pilots calls our attention to the need for participants to have developed a long-term strategy around resources being allocated to their portfolio of projects – in terms of time horizon of commitment, and human and financial capital. Repeated forms of participation in demonstration pilots create the foundation for developing competencies and building a trust network of collaborators. These collaborators can compete in the global marketplace for sustainable energy innovation as opposed to following the one-off trial-and-error experimental learning strategy that, at best, may result in intellectual property gains.

This chapter has discussed organizational learning and how essential it is for the creation of impact at the enterprise level. Starting from innovative companies such as Japan's Eisai, we have looked at the importance of long-term value creation and how both learning and unlearning are essential stages for an organization to adopt and sustain responsible business practices.

What, then, are the tools that will help companies achieve their sustainability ambitions?

Is having access to more data the answer?

THE LEARNING JOURNEY – ORGANIZATIONAL LEARNING FOR IMPACT

For businesses to clearly articulate impact objectives and successfully deliver them as part of their sustainability journey, they are called to reflect on the learning processes that affect the organization. The following checklist is organized in inquiry form to help articulate the most suitable organizational learning path for a business of your choice.

- Which are the learning frameworks that most characterize the culture of the organization under review? Do they trigger a culture of accountability and performance against the company's defined sustainability targets?
- Deriving your intuition from the Yanagi Model as it relates to human capital variables, identify the KPIs – quantitative and qualitative – that are most aligned with a participatory learning environment. How would you recommend amending those that seem to lack alignment through an analytical framework?
- Unlearning mechanisms may stir innovation-oriented organizations to adopt a culture of sense-making while also allowing for structural time delays in scaling sustainable outcomes. Draw an intuitive map of stakeholders to illustrate the mental models that seem to characterize their response to impact goals. Point out the areas (product/function/geography) where short-term resource allocation decisions seem to create roadblocks to support impact-oriented organizational decisions. How do time delays favor or slow down organizational effectiveness of sustainable projects?

TECHNICAL NOTE – THE YANAGI MODEL

First introduced in 2018 by Dr. Yanagi, the model aims to identify the relationship between the value of corporate investments in intangible assets (e.g., environmentally and socially relevant assets, or non-financial capital) and corporate value as defined by the traditional shareholder's metric of share price-to-book value ratio (PBR).

By leveraging the empirical evidence from Eisai Co., Ltd., Dr. Yanagi's findings underscore the statistical relevance of non-financial capital investments (such as human capital and metrics of employee wellbeing) toward positive contribution to future profitability of the organization in the medium to longer term.

By employing multiple-regression analyses of Eisai's KPIs (88 factors) over a nearly 30-year history of PBR data and the Digital ESG Platform by ABeam Consulting, Dr. Yanagi observed the time lag of corporate investments in non-financial metrics of environmental and social relevance to Eisai. There is statistical significance ranging up to a ten-year time lag for 20 out of the 88 KPIs analyzed. Specifically, the model findings reinforce the connection between organizational learning involving non-financial measures of impact and the ability of organizations utilizing analytical frameworks to better align strategic planning with sustainability objectives. The regression analyses conducted by Dr. Yanagi have been translated in sensitivity analyses to drive allocation of investments in the wellbeing of the labor force and to strengthen Eisai's intellectual capital through increased R&D investments (Figure 2.5).

The Yanagi Model builds on prior studies by its author, which bring quantitative evidence to the definition of shareholder value as the sum of a metric of Market Value Added (MVA) plus the corresponding accounting book value of shareholders' equity (BV). In addition to the development of the pillars of empirical research supporting the introduction of the MVA and the alignment of non-financial capital metrics with the framework set forth by the International Integrated Reporting Council (IIRC), Dr. Yanagi provides qualitative evidence in support of the model's empirical findings drawn by extensive multistakeholder engagement surveys.

As Global CFO of Eisai, Co., Dr. Yanagi continues to promote the need to embed an organizational learning roadmap within the global finance organization at the company, as well as across the full spectrum of personnel. The direct involvement of unionized labor in the discussion of human capital investments and direct ties with longer term value creation in financial terms emphasizes the importance of a participatory environment in driving organization learning and sustaining the longer-term focus of business decision-making to achieve impact-oriented objectives.

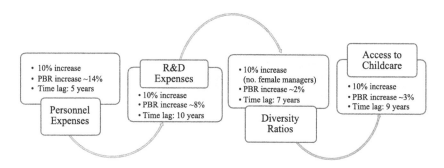

FIGURE 2.5   The Yanagi Model: sensitivity analysis with observed time delay. Author referenced Eisai (2021). Note: the "Eisai Integrated Report", 55, was partially updated by Dr. Yanagi as of May (2021).

NOTES

1 The author thanks the attendees to the Women in Finance Conference and Awards 2020 – Canada and the Bonhill Group UK for the roundtable discussions and group activities that have inspired the author's research on the topic of organizational learning curves for impact-oriented businesses and the roadmap presented in this chapter.
2 Note on the use of the software Mental Modeler: it helps illustrating both factors affecting the decision-making context and their relationships through an influence map.
3 Profile from RepRisk® Reputational Risk and ESG Due Diligence Database as of January 2021.
4 The author thanks the participants to the 2021 international conference on cooperation in the hydrogen energy industry which was held in Chengdu, China, in May 2021, the Green and Smart Energy Organization, the Sichuan Provincial Bureau of Economic Cooperation, the China Council for the Promotion of International Trade and the China Construction Bank (Sichuan Branch).

REFERENCES

Akgün, A. E., J. C. Byrne, G. S. Lynn, and H. Keskin. 2007. "Organizational Unlearning as Changes in Beliefs and Routines in Organizations." *Journal of Organizational Change Management* 20 (6): 794–812. doi:10.1108/09534810710831028.

Bossink, B. 2020. "Learning Strategies in Sustainable Energy Demonstration Projects: What Organizations Learn from Sustainable Energy Demonstrations." *Renewable and Sustainable Energy Reviews* 131. doi:10.1016/j.rser.2020.110025.

Denrell, J., C. Fang, and D. A. Levinthal. 2004. "From T-Mazes to Labyrinths: Learning from Model-Based Feedback." *Management Science* 50 (10). doi:10.1287/mnsc.1040.0271.

Douglas, E. M., S. A. Wheeler, D. J. Smith, I. C. Overton, S. A. Gray, T. M. Doody, and N. D. Crossman. 2016. "Using Mental-Modelling to Explore How Irrigators in the Murray–Darling Basin Make Water-Use Decisions." *Journal of Hydrology: Regional Studies* 6: 1–12. doi:10.1016/j.ejrh.2016.01.035.

Eisai Co., Ltd. 2020. "Eisai Integrated Report". 26 October: 14. https://www.eisai.com/ir/library/annual/pdf/epdf2020ir.pdf.

Gray, S. A., S. Gray, L. J. Cox, and S. Henly-Shepard. 2013. "Mental Modeler: A Fuzzy-Logic Cognitive Mapping Modeling Tool for Adaptive Environmental Management." *2013 46th Hawaii International Conference on System Sciences*: 965–973. doi:10.1109/HICSS.2013.399.

Guerrero, A. M., N. A. Jones, H. Ross, M. Virah-Sawmy, and D. Biggs. 2021. "What Influences and Inhibits Reduction of Deforestation in the Soy Supply Chain? A Mental Model Perspective." *Environmental Science & Policy* 115: 125–132. doi:10.1016/j.envsci.2020.10.016.

Heiskanen, E., K. Hyvönen, S. Laakso, P. Laitila, K. Matschoss, and I. Mikkonen. 2017. "Adoption and Use of Low-Carbon Technologies: Lessons from 100 Finnish Pilot Studies, Field Experiments and Demonstrations." *Sustainability* 9 (5): 1–20.

Kolkman, M. J., M. Kok, and A. van der Veen. 2005. "Mental Model Mapping as a New Tool to Analyse the Use of Information in Decision Making in Integrated Water Management." *Physics and Chemistry of the Earth, Parts A/B/C* 30 (4–5): 317–332. doi:10.1016/j.pce.2005.01.002.

Maitlis, S., and M. Christianson. 2017. "Sensemaking in Organizations: Taking Stock and Moving Forward." *Academy of Management Annuals* 8 (1). doi:10.5465/19416520. 2014.873177.

Michael, D. N. 1985. *On Learning to Plan and Planning to Learn*. San Francisco, CA: Jossey-Bass.

Moon, K., A. Guerrero, V. Adams, D. Biggs, D. Blackman, L., Craven, H. Dickinson, and H. Ross. 2019. "Mental Models for Conservation Research and Practice." *Conservation Letters*. doi:10.1111/conl.12642.

Owen, R., P. Macnaghten, and J. Stilgoe. 2012. "Responsible Research and Innovation: From Science in Society to Science for Society, with Society." *Science and Public Policy* 39 (6): 751–760. doi:10.1093/scipol/scs093.

Rahmandad, H., N. Repenning, and J. Sterman. 2009. "Effects of Feedback Delay on Learning." *System Dynamics Review* 25 (4): 309–338. doi:10.1002/sdr.427.

Rapp, D. 2005. "Mental Models: Theoretical Issues for Visualizations in Science Education." *Visualization in Science Education* 1: 43–60. doi:10.1007/1-4020-3613-2_4.

Runyan, C. W., and J. Stehm. 2020. "Deforestation: Drivers, Implications, and Policy Responses." *Oxford Research Encyclopedia of Environmental Science* (July 30). doi:10.1093/acrefore/9780199389414.013.669.

Schein, E. H. 1985. *Organizational Culture and Leadership*. San Francisco, CA: Jossey-Bass.

Schein, E. H. 1992. *How Can Organizations Learn Faster? The Problem of Entering the Green Room*. MIT Sloan School of Management, WP# 3409-92. https://dspace.mit.edu/bitstream/handle/1721.1/2399/SWP-3409-45882883.pdf.

Schlosser, C. A., K. Strzepek, X. Gao, C. Fant, É. Blanc, S. Paltsev, H. Jacoby, J. Reilly, and A. Gueneau. 2014. "The Future of Global Water Stress: An Integrated Assessment." *Earth's Future*, 2 (8): 341–361. doi:10.1002/2014EF000238.

Senge, P. M. 1990. *The Fifth Discipline*. New York: Doubleday.

Van Geenhuizen, M., and Q. Ye. 2014. "Responsible Innovators: Open Networks on The way to Sustainability Transitions." *Technological Forecasting and Social Change* 87: 28–40. doi:10.1016/j.techfore.2014.06.001.

Yanagi, R. 2018. *Corporate Governance and Value Creation in Japan*. Singapore: Springer Nature.

Yanagi, R., and N. Michels-Kim. 2021. "Eisai's ESG Investments." *Strategic Finance* (May). https://sfmagazine.com/post-entry/may-2021-eisais-esg-investments/.

Yanagi, R., and S. Sugimori. 2021. "Delayed Penetration Effect of ESG KPIs on PBR and Disclosure in Integrated Report." *Kigyo Kaikei* (January): 112–120.

Yang, K. P., C. Chou, and Y. J. Chiu. 2014. "How Unlearning Affects Radical Innovation: The Dynamics of Social Capital and Slack Resources." *Technological Forecasting and Social Change* 87: 152–163. doi:10.1016/j.techfore.2013.12.014.

3 Are Novel Datasets the Solution?

Organizations learn differently and often follow what can be called "group neural pathways" when introducing new information or tools for the purpose of decision-making. These established group patterns are based on acquired knowledge and conditioned by past outcomes. They may prevent the same group from experimenting further when it comes to drawing insights from novel datasets and borrowing tools from other industries.

Culture – "the way things get done" within an organization – contributes to that. It crystallizes learning by area of domain expertise at the individual level and may ultimately slow down adoption of new inputs in the decision-making process. Structural change and systemic disruptions can change that. It is at these times that organizational learning can move into a recovery or growth mode that in fact is adaptive to a new context. This is what happened during the COVID-19 pandemic. Such times of disruption and change can also lead to new measures of organizational impact and a new organizational lens that prioritizes the value delivered by collective actions.

Inherently, scientists and organizational leaders that seek to integrate sustainability objectives (and impact outcomes) in their day-to-day decisions become more aware of the functional areas where embedding the evaluation of impact may yield both economic and societal benefits. How can they leverage novel sources of data to alter group thinking and enable a breed of decision-making that focuses on measurable results?

Whether for planning budgets, launching products or hiring talent, most datasets built on Environmental, Social and Governance dimensions have already been introducing new sustainability parameters. Those parameters are nondirectly quantifiable and come in metric units different from those that organizations have traditionally used to inform decision-making. Examples include live data streams collected through remote sensing and GPS trackers that report on forest management and energy security through geospatial analytics; survey indicators that gauge employee engagement and assess the wellbeing of the workforce; and compliance metrics that meet industry standards addressing ecosystem biodiversity and environmental degradation, to mention just a few.

In addition to novel datasets, organizations have begun to explore where traditional sources of data, whose functionality is being extended from one impact area to another, apply to a variety of emerging use cases. An example is adapting customer-oriented product or service features to address in-house talent retention needs or expand workforce retirement benefits and employee experience

DOI: 10.1201/9781003212225-4

platforms. By delivering indicators of organizational belonging, a professional objective can be aligned with personal values in the workplace.

Businesses learning about sustainability – either for their internal orientation or to equip their teams with expertise in bringing to market products and services for environmental and societal conscious consumers – find two common roadblocks when attempting to merge novel data sources with traditional metrics:

1. the uneasy access to freely available impact data in an open architecture setting as limited connectivity slows down analytics and visualization through application programming interfaces (APIs), and
2. alternative data sources lack transparency in terms of collection methodologies. Verification and assurance processes are still in their infant stages, which prevents a robust integration of potential insights for the purpose of decision-making.

The regulatory frameworks and policy actions emerging as a result of the rebuilding efforts post-COVID-19 may create compelling conditions for public and private sector participants to overcome these challenges. Securing direct access to comparable data feeds and establishing trusted environments to connect openly available sources of alternative information is a key step in enabling impact-oriented decisions, while also fostering collaborative learning and strengthening peer networks. Where open access is embraced, visualization techniques such as digital prototyping can be employed to build user-defined scenarios of organizational commitments and envision future outcomes.

Let us look at an example from the public sector. By adopting the concept of a "digital twin" to aid decision-making, an increasing number of governments and municipalities have been piloting digital prototyping techniques for the built environment. The purpose of these tools is to enable visualization and more directly frame the what-if scenarios in the context of policy decisions. From circular economy solutions, to reforestation needs that address environmental commitments such as tracking net zero carbon emissions, to disaster management of public infrastructure in smart cities, the applications of digital prototyping are far from exhausted.

The City of Philadelphia in the US offers an early example of how aligning long-dated sustainability commitments requires an ongoing effort in digitizing relevant data and planning a more sustainable urban environment to meet those commitments. In early 2021, Philadelphia released its 2050 carbon neutrality commitment, which included the publication of its first Climate Action Playbook. The municipality also decided to strengthen its expertise by establishing a Technical Panel on Climate Science and Research to advance resilience planning and deliver its 2050 goals.

Scientific methods offer well-developed insights and tangible evidence of the tools needed to combat a climate crisis, but it is businesses and governments that are tasked with building data-informed resilience-planning scenarios. These may range from actions that aim at creating positive externalities – fostering a transition to a greener built environment – and others that focus entirely on reducing the negative impact of climate deterioration. In the case of the built ecosystem, digital prototyping can be an essential tool in identifying both positive and negative externalities. In the case of urban ecosystems like the City of Philadelphia, which have added nature solutions such as reforestation to the mix of policy actions, the direct application of digital prototyping enables municipalities to redesign urban landscaping needs. For example, it is possible to quantify analytically the optimal tree coverage required to achieve an air quality target.

Incorporating the digital insights of piloting efforts may yield faster adoption of alternative sources of data. It may also provide the foundation to expand its use from individual municipality and state to private sector organizations, and guide government policies.

ALTERNATIVE SOURCES OF SUSTAINABILITY DATA

Commitments to adopt sustainability metrics or hefty multiyear targets continue to pose challenges for businesses. How can they implement science-based planning with an eye on the operational monitoring, tracking and executing of data-oriented roadmaps to achieve targeted impact outcomes? I use the term "alternative data" to define novel forms of data to fill in the "value" gaps in traditional metrics and/or provide additional insights that may bridge the knowledge gap. Anecdotal evidence suggests that there are over 600 fields currently in use to define environmental or societal outcomes, many qualitative and policy-oriented, others more readily quantifiable. As the sustainability narrative becomes an integral part of corporate jargon and reporting of data and policies proliferates, one can only expect that number to increase dramatically. Mindful of that, the fast growth of trusted digital platforms to store, analyze and visualize sustainability data is rooted in two primary elements:

1. impact outcomes are becoming part of the value delivered by an enterprise, and
2. impact outcomes are directly associated with a wide range of stakeholders whose input historically may have lacked a conduit to influence strategic business planning.

In essence, science-based sustainability commitments are supportive of the concept of stakeholder capitalism. This is the notion that corporations are called to create value for a set of beneficiaries beyond those driven by financial returns (shareholders or lenders) to include employees, customers, consumers, supply chain providers and other vendors.

Intuitively, expanding the information that a company taps for operational decisions to include alternative sources of sustainability data that reflect those broader stakeholder categories would enable a direct mapping of their insights to the day-to-day aspects of a business. Aspects that are beyond what traditional financial accounting or procurement analyses and purchasing decisions may have historically relied on. For example, alerts related to controversial business activities that vendors may have been involved in can be embedded in the cost-benefit analysis run by procurement analysts and include reviews of how the severity of such controversies may hinder the continuity of the purchasing cycle.

In light of controversies involving human rights, bribery and potential corporate complacency toward environmental damages, investors continue to place significant pressure on companies to report their own operational readiness. In fact, mandatory due diligence of controversial business activities is increasingly part of standard contractual obligations to protect the business should any wrongdoing by a partner organization emerge.

Integrating impact as a value driver while also opening the dialogue with a wide range of stakeholders is the multidimensional problem addressed in the science of sustainability. The solution is hard to design and far harder to calibrate to business objectives that have prioritized near-term financial outcomes. Introducing novel data points or even embarking in data collection of environmental and social dimensions that may not necessarily "fit" in one business area but rather cut across functional silos, product lines and geographies poses a unique challenge. That challenge is operating a trust network and opening access to all interested stakeholder groups as a way that promotes cross-validation of information exchange and early feedback.

The ecosystem in which alternative sources of data reside – where they are stored, populated and aggregated in preparation for visualization and prototyping exercises – is as important as the validation of data sources and the inevitable cleaning of data that occurs when data is collected. Mapping stakeholders to data streams and enabling early feedback involves designing business-wide open data initiatives. This promotes the formation of peer networks and encourages faster learning on the use of alternative variables in decision-making – a catchall for unveiling both pitfalls and opportunities. The open environment setting also means that, when it comes to business decisions, financial stakeholders are likely to hold a more diluted weight.

At this stage in the impact learning process, there is no need to rush to integrate these variables with pre-existing metrics of economic value. They need to first provide insights on customer positioning and prototyping new scenarios. The residual question of value delivered or contributed should be left to a later time. In fact, if sources of alternative data are introduced before establishing which internal infrastructures will benefit from their insights, efforts may risk becoming merely a regulatory compliance exercise – an obligation by the organization to map and maintain metrics of responsible business conduct instead of leveraging them as value enhancers and drivers of new market opportunities.

Japanese biotechnology innovator Eisai again provides an example when we look at its learning journey through the lenses of its CFO, Dr. Yanagi. Mapping financial resilience and financial growth to human capital was addressed by the pioneering process of utilizing social metrics such as gender, employee wellbeing and engagement, less as control variables in a sample set, but more as primary variables in a principal component analysis of direct contributors to long-term enterprise value. They were not objectives to optimize against, but variables (in this case, social in nature) directly correlated to growing enterprise value in economic terms.

A question that frequently emerges is: Why is it so important to lay out the integration of alternative sources of sustainability information from the perspective of the learning journey of an organization?

The answer is that collaboration tools such as digital prototyping aim at addressing manageable risks as well as providing the foundation for decision makers to steer away from unmanageable scenarios before committing entire organizations to unattainable goals – whether sustainability-oriented or not.

When elevating multiyear sustainability commitments such as reductions in carbon emissions or increased access and inclusion in the customer outreach plans of an enterprise, tangible strategic objectives and cross-functional priorities need to be established. By introducing non-traditional sources of information and creating a reliable data inventory of existing as well as novel data sources, organizations realize fairly quickly the need to develop foundational infrastructure in support of open sharing. This is done either by breaking internal functional or geographical silos, or by creating a trusted peer environment. That environment must be supported by digital capabilities to address target impact areas and build the necessary buy-in for a data-driven, open access effort across the organization.

BUILDING A TRUST NETWORK TO MEET THE SDGs

Environmental and social commitments are increasingly being expressed in terms of impact outcomes to a wide range of stakeholders, not just as short-term economic exposures of shareholders and others that hold financial interests in an enterprise. This makes the process of mapping data streams and insights to the relevant stakeholder groups an essential step to capture impact-oriented outcomes. When it comes to large-scale open architecture projects, the role of the Open Data Institute (ODI) in the UK is regarded as a pioneer in the field.[1]

> We observed that early on, people leading new data access initiatives often need help understanding what data infrastructure exists in their data ecosystem, and how to access it. They also needed support in effectively engaging and utilizing stakeholders from across the data ecosystem, especially those from different sectors.
>
> *(ODI 2021a)*

From guidance on how to best identify the relevant stakeholder categories to the creation of data inventories, the ODI provides rich examples of how, when businesses start addressing their internal user data needs while also planning for an operating environment built on trust and shared learning as the project evolves, they can leverage decision-useful platforms to house feedback from all stakeholders. Specifically, by employing ODI's practical guidance on building open data ecosystems, organizations are more likely to focus on the insights contributed by each new piece of information. The introduction of non-traditional sources of data that address broadly defined sustainability areas for impact has the potential for large-scale integration and deeper adoption among stakeholders.

Long-time followers of the impact dialogue among public sector participants and the international NGO community will not be surprised to observe that the UN Sustainable Development Goals (the SDGs) have amplified the early impact narrative. Since their launch in 2015, the SDGs have channeled impact evaluation and measurement efforts according to 17 areas of focus. There are over 230 indicators, which target issues such as alleviation of poverty, socio-economic wellbeing and environmental degradation, to name just a few. Many of these goals may come across as aspirational. But these goal-setting activities have prompted the need to look for alternative forms of information – to assess, monitor, verify and compare pathways and progress toward achieving those goals. Historically, traditional data to build SDG-aligned indicators has been mostly sourced from national statistical offices, governments and international organizations. This represents the foundation of early attempts to create a more systematic way of reporting and monitoring progress. The cost of aggregating standardized data streams for traditional indicators is not inexpensive. It is mostly derived from surveys of hundreds of millions of people, leaving more affluent countries with a greater data capacity footprint and, therefore, an advantage in terms of data coverage, availability and robustness of inputs.

Maintained by the United Nations Statistics Division, the Open SDG Data Hub is an example of an open platform that has made it easy to compare coverage and accessibility of SDG-relevant data and statistics by geographic region in a disaggregated way. It does this by relying on geospatial databases. A closer assessment of the data inventory and its update cycle reveals a clear message of the correlation between national statistical capacity and each region's openness to support data availability on a global scale. It also contributes by allocating resources to define relevant SDG indicators and source information in a transparent fashion[2] (Figure 3.1).

THE VALUE OF NON-TRADITIONAL DATA BEYOND THE SDGs

In 2015, I started conducting company-level assessments and interviews about organizations' strategic planning and the integration of sustainability-oriented business scenarios. At this time, the utilization of non-traditional data beyond impact evaluation of the newly launched SDG targets was minimal. Now, as businesses rush to adopt science-based commitments, in particular for environmental pathways to 2030, it is easier to look back and notice anecdotally how the lack of collaborative efforts has amplified the divide between sustainability "leaders" and "laggards." This has created a void that may severely affect the less-resourced organizations in terms of capital and talent. In addition, the efforts of individual organizations to define and build their own sustainability data ecosystem and chart their own journey into data-informed sustainable practices stress the urgency of casting a wide net to non-traditional data beyond its use for impact reporting through the SDGs.

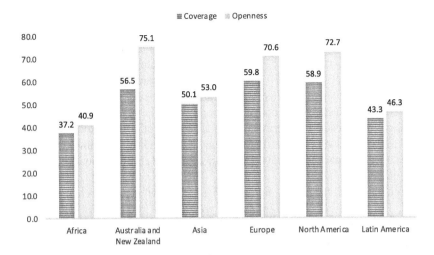

FIGURE 3.1 The realm of traditional data: coverage and openness (2020). Note: Assessment of coverage and openness of official statistics represents the degree of accessibility of the UN Sustainable Development Goals statistics per geographic region as maintained by official statistical offices on a scale 1:100, in increasing order. Data: Open Data Watch – Open Data Inventory http://www.opendatawatch.com. Update as of June 2021. Author also referenced Fritz (2020).

In my interviews (conducted between 2015 and 2020) it emerged that most organizations are still deciding which SDG targets to adopt and why, and when to start the clock on a progress timeline. In fact, the idea of establishing dynamic targets is not even being talked about by organizations. The rush is on to develop an organizational roadmap to establish quick wins and build trust among peers, customers and employees, while growing the organizational readiness to embrace a well-functioning and measurable impact journey.

The value of non-traditional data for strategic decision-making has become more apparent to private and public sector participants. This is due to the emergence of open innovation initiatives and platforms that allow companies to tap the knowledge and experience of experts outside their traditional domains and leverage digital technologies to boost organizational learning and shared best practices.

One example is the recent use of big data in handling the analyses of non-traditional metrics in the global tourism industry. In 2005, 10 years prior to the launch of the SDGs, the World Tourism Organization, in cooperation with the United Nations Environmental Programme, argued in their *Making Tourism More Sustainable: A Guide for Policy Makers* that "the environmental, economic, and socio-cultural aspects of tourism development and a suitable balance must be established between these three dimensions to guarantee its long-term sustainability" (World Tourism Organization 2005, 11).

While allowing for quick experimentation and almost immediate feedback from customers, the automated manipulation of large datasets allows us to gather insights that can directly improve daily operations through efficiencies while also building a novel approach to customer segmentation. Pooling external knowledge through open innovation platforms and leveraging digital technologies to promote external stakeholder engagement – which in the case of the tourism industry comprises both resident and tourist populations – has been proven to greatly enhance visitors' experiences. Analyses of preferences are conducted with social analytics that enable visualization and mapping of emerging consumer preferences, such as those related to sustainable consumption and trends in agri-business conditions by geographic location. Nevertheless, the scalability of social analytics remains constrained by issues associated with open data availability plus ambivalent sentiment towards legal and ethical considerations around platforms powered by artificial intelligence.

Digitization is also promoting a significant redesign of supply chain relationships. The traditional cost optimization paradigm of procurement decisions is being challenged by the need for transparent transactions and regulatory compliance with respect to environmental and social impact. Sourcing networks built on "trust" rely increasingly on the use of digital ledger technologies such as blockchain to ensure supply chain transactions reflect product provenance, authenticity and legitimacy as vital elements of the procurement cycle.

When coupled with digital mapping and terrestrial sensors, these platforms have achieved a high level of sophistication in just a handful of years. They have also opened the door for global businesses to adopt sustainable procurement policies and increasingly move purchasing decisions towards building transparent procurement networks by employing registered product-level barcodes and digitally storing product characteristics. Such characteristics can include provenance and country of origin and, in the not-so-distant future, carbon emissions associated with production and distribution networks. Comparability and transparency of product-level data regarding environmental and social impact are in their early stage, but digital ledger applications are starting to redefine the governance setting where impact data can be freely exchanged. This creates a permanent record of its evolution while improving the visibility and legitimacy of transactions.

The "trust" of blockchain-enabled exchanges is potentially revolutionizing. Impact-oriented organizations looking to leverage their digital infrastructure can grow trust in the social and environmental targets they have adopted by tapping into blockchain platforms.

TRUST AS AN IMPACT VARIABLE[3]

Over the past decade, the data ecosystem has moved from preserving privacy across its multiple facets – whether regulatory, consumer-driven, or cyber risk focused – to prioritizing transparency and the ability of organizations to build a trust environment when evaluating data insights in their global business and stakeholder networks.

In their 2021 report for the ODI, *The Economic Impact of Trust in Data Ecosystems*, the research team at consultancy Frontier Economics analyzes how trust correlates with the level of data flow. Increased trustworthiness of the data ecosystem (e.g., in data collection, sharing and use) generates an increase in its economic and social value as a result of improved public and private sector decision-making and a favorable backdrop for innovation.

Why is sense-making so important when addressing the emergence of novel datasets and what to do with them? Because technology adoption and digital innovation require significant capital investment and the ability to bring suppliers – and their suppliers – into the equation. Technological sense-making for impact is key to building an ecosystem of trust in onboarding non-traditional sources of information and data for institutionalized transactions that already have a long history, such as purchasing decisions and responsible sourcing. How can we best address that? (See Technical Note for an overview of building trust environments and the selection of a digital ledger platform for supply chain due diligence.)

In its most basic application between a buyer and a seller, issuing a bloc directly affects marketplaces where the origin of a product or manufactured good is a critical component of the decision to buy. The bloc in itself replaces the trust building role that financial intermediaries may have held in the past and the stamped documentation that would seal the final purchase. While multitier supply chains are a more complex problem, a common feature of building trust has historically been relational in nature. Long-term financial commitments between counterparties in the exchange lay the foundation required for building longer term trust among supply chain participants (i.e., financial obligations and the follow through on financial creditworthiness as a trust building exercise). As blockchain exchanges potentially protect from fraud, tampering and cybercrime, data integrity is in fact regarded as a key attribute of blockchain-based ledgers. Information stored in the ledger is immutable and joined in a linear chain, and it is therefore auditable (especially in ledgers that are organized in a permissioned environment).

The efficiency gains of implementing ledger-prone supply chains are easy to assess – time spent in transit and tracking an in-transit good is reduced, and logistics management is more direct. But it also plays a direct role in matching financial obligations with receivables at various levels of reported inventory. This in turn creates a more accurate assessment of demand forecasts, produces a more balanced order mechanism and optimizes cargo shipping, while also checking the box on responsible sourcing. Looking forward, a more streamlined multi-vendor comparison mechanism among blocs for each good could help organizations make better decisions in terms of the underlying impact they are looking to make. Either societal or environmental benefits can be attached to it, including, for example, circular economy gains in waste reduction and better monitoring of potential spoilage and overall safety of goods.

As companies gain deeper familiarity with non-traditional sources of data and partner with a diverse spectrum of information providers for insights, they can help create a trust environment spanning from the development of technical tools to the development of forward-looking metrics of organizational impact.

How can organizations measure their progress toward impact goals?

How do they redefine the allocation of talent and resources to fast-track progress?

THE LEARNING JOURNEY – ARE NOVEL DATASETS THE SOLUTION?

A decision matrix to guide the integration of alternative data sources is a necessary tool to ensure that organizational focus on key impact outcomes translates to targeted data collection and analyses. The following checklist is organized in a step-by-step format to help the reader map the environmental and social targets of a business to its data needs and the stakeholders involved in the evaluation of the insights generated by ESG data.

* Enlist the organizational commitments (sustainability targets, impact areas or ESG goals of the business). Distinguish between formal/ informal commitments, any time horizon for progress monitoring, and who are the internal/external stakeholders involved ("who cares about the goals?").
* Enlist existing data sources related to impact areas within the relevant Environmental (E), Social (S), and Corporate Governance (G) categories. Answer the question: are E, S, G metrics already available? Assess whether they are qualitative/quantitative metrics; internal/external sources of information.
* Enlist stakeholder groups associated with existing ESG data fields. Are they also internal owners of the information?
* Focus on internal and external stakeholder groups. Anyone missing from existing pool/consumers of readily available ESG metrics; how to reach them and what metrics are they looking for?

Note: This checklist can be used upside down to identify key sustainability commitments that an organization needs to make, given the stakeholder groups the business seeks to reach or establish or strengthen an existing relationship with.

TECHNICAL NOTE – BUILDING A TRUST ENVIRONMENT WITH A DIGITAL LEDGER PROTOCOL

Minespider operates a raw material supply chain infrastructure platform that uses digital ledger technology to trace minerals and other commodities along the entire supply chain. The companies that use the platform want to understand their supply chain exposure to major controversial activities such as forced, bonded or child labor, corruption and counterfeiting in their global supply chains, and various degrees of separation from a company's purchasing decisions. In most cases, companies employ platforms such as Minespider's to lower their reputational risk and mounting pressures from a variety of stakeholders, including responsible investors and the general public.

The Minespider Protocol offers companies a way to understand where those issues reside in their network as a complement to supplier-level questionnaires, which have been historically used by companies to carry out basic due diligence of their third-party providers. More often than not, these questionnaires are neither timely nor effective in collecting point-to-point observations, nor can they be leveraged fully to comply with major norms and regulatory asks. Through pilot programs that trace the supply chains of metals in select countries, the platform has been able to more efficiently collect information needed to unveil weaknesses in suppliers. This enables companies to strengthen compliance monitoring activities for their purchased materials. The core tenets of the protocol underscore the importance of curating a trust environment at each step of the data collection exercise to ensure trust is maintained during the due diligence process.

Looking forward, an opportunity to evolve the service of blockchain protocols and platforms such as Minespider is likely to emerge with the creation of a decommoditized market for sourced materials. In fact, buyers will be looking to streamline purchase decisions by interacting with counterparties that are willing and able to share premium characteristics of a commodity, including traceability, quality and control features, against environmental and social controversies, and to ensure the sourcing of materials rewards sustainable commodities and incentivizes sustainable procurement practices.

TABLE 3.1
Profile of a Digital Ledger Protocol for Impact

Profile of Digital Ledger Protocol for Impact	Hierarchy of Trust Mapped to the Digital Ledger Characteristics
Open Source	Prioritization of active and positive impact outcomes by the business is consistently communicated.
Data Agnostic	Data quality and coverage is fit for purpose. The business ensures quality and accuracy checks through appropriate data governance infrastructure.
Third-Party Audited	Governance and strategic oversight of all data sources and measurement of outcomes generated is subject to internal and external monitoring.
Decentralized	Data privacy and transparency are prioritized through a design of permissioned blockchain.
Interoperable	Traceability is ensured across entire supply chain by leveraging modular design of digital ledger.

Source: Author's intuition. Hierarchy of trust column adapted from the ODI'S *Trustworthy Data Stewardship Guidebook* (2021). Profile of digital ledger protocol for impact makes reference to the Minespider Protocol v.0.36.

NOTES

1 The author thanks Stuart Coleman, Director of Learning and Business Development at the ODI, for his insightful introduction to data strategies and the ecosystem approach advocated by the ODI.
2 The author provided her expert opinion to the Inter-Agency Task Force on Financing for Development of the UN Department of Economic and Social Affairs in the area of ESG investing and the use of unharmonized and uneven information sets provided by businesses in their sustainability reports (2018–2019).
3 The author highlights the proceedings of the 2019 Global GRC (Governance, Risk and Compliance) Summit hosted by MetricStream in Baltimore, USA under the theme "Perform with Integrity" and her expert talk on "Refocusing on Reputational Risk in the Risk Framework."

REFERENCES

The City of Philadelphia, Office of Sustainability. 2021. *Philadelphia Climate Action Playbook.* https://www.phila.gov/media/20210113125627/Philadelphia-Climate-Action-Playbook.pdf.

Fritz, S., L. See, T. Carlson, M. M. Haklay, J. L. Oliver, D. Fraisl, R. Mondardini, M. Brocklehurst, L. A. Shanley, S. Schade, and U. When. 2019. "Citizen Science and the United Nations Sustainable Development Goals." *Nat Sustain* 2: 922–930. doi:10.1038/s41893-019-0390-3.

Frontier Economics. 2021. "Economic Impact of Trust in Data Ecosystems." Report Prepared for the ODI. https://theodi.org/article/the-economic-impact-of-trust-in-data-ecosystems-frontier-economics-for-the-odi-report/.

Herzog, T. 2019. "New Resources for ESG Data and Investors." *World Bank Data Blog*, October 2019. https://blogs.worldbank.org/opendata/new-resources-sovereign-esg-data-and-investors.

Open Data Institute. 2019a. "A Match Made in Heaven: How 'Digital Twins' Can Help Bring a Better Built Environment." February 2019. https://theodi.org/article/a-match-made-in-heaven-how-digital-twins-can-help-bring-a-better-built-environment/.

Open Data Institute. 2019b. "Data Ecosystem Mapping Tool." May 2019. https://theodi.org/article/data-ecosystem-mapping-tool/.

Open Data Institute. 2020. "Digital Twins – Virtual Versions of Real-World Assets." December 2020. https://theodi.org/article/digital-twins-virtual-versions-of-real-world-assets/.

Open Data Institute. 2021a. "Exploring How New Data Access Initiatives Approach Date Landscaping. January 2021. https://theodi.org/article/exploring-how-new-data-access-initiatives-approach-data-landscaping/.

Open Data Institute. 2021b. *Trustworthy Data Stewardship Guidebook.* https://theodi.org/article/introducing-the-odi-trustworthy-data-stewardship-guidebook/.

Open Data Watch. "Open Data Inventory." http://www.opendatawatch.com.

Schmidt, S. G., and S. M. Wagner. 2019. "Blockchain and Supply Chain Relations: A Transaction Cost Theory Perspective." *Journal of Purchasing and Supply Management* 25 (4): 100552. doi:10.1016/j.pursup.2019.100552.

SDG Knowledge Hub, Policy Brief. 2020. "Tracking the Trackers." 23 July 2020. http://sdg.iisd.org/commentary/policy-briefs/tracking-the-trackers-sdg-data-sources-at-year-five/.

UN Department of Economic and Social Affairs, Open SDG Data Hub. https://unstats-undesa.opendata.arcgis.com/.

Wang, Y., M. Singgih, J. Wang, and M. Rit. 2019. "Making Sense of Blockchain Technology: How Will It Transform Supply Chains?" *International Journal of Production Economics* 19: 221–236. doi:10.1016/j.ijpe.2019.02.002.

Williams, N. 2018. "Minespider: Protocol for Due Diligence in the Raw Materials Supply Chain." White paper, v. 0.36. https://uploads-ssl.webflow.com/6098de8910ab 20fb71ac62b9/60cb022e84f17b18e4a46575_Minespider%20v0.4%20-%20 Light%20Paper.pdf.

World Tourism Organization; United Nations Environmental Programme. 2005. *Making Tourism More Sustainable: A Guide for Policy Makers.* Madrid: World Tourism Organization Publications. https://wedocs.unep.org/handle/20.500.11822/8741.

4 Taking Impact a Step Forward

So far, we have focused on organizational learning as a way to retool businesses from within and build resilience by embedding sustainability objectives in decision-making. The way a business manages this learning journey and sets specific impact targets and progress timelines is likely to reshape its modus operandi from a risk management, strategic planning and innovation perspective. While grassroots initiatives may spark broad-scale participation and foster cooperation across product and functional areas, in over two decades of research I have found that successful organizations focus on integrating these insights in their learning roadmap in a way that supports, yet challenges, existing metrics of success.

The emergence of global sustainability standards and standardized forms of financial accounting for sustainability metrics have helped bridge the gap between corporate commitments and the operational needs of businesses. Nevertheless, because of the inevitable time delays in translating metrics into corporate value, many corporate efforts are just sporadic attempts to contextualize environmental and social metrics. Near-term insights are assigned to operational business scenarios that may be out of sync with strategic planning decisions to sustain future business growth.

One of the key challenges in embedding impact-oriented data is the difficulty in creating effective reporting. This reporting needs to be just as bullet-proof and relevant for external communication (to investors, consumers, potential future talent, communities, etc.) as other public statements that a business makes (so subject to litigation and transparency claims). There continue to be issues surrounding the voluntary reporting of information. The gap in standardized reporting practices will most likely take several years to close before meaningful adoption of standards by organizations more broadly. For example, it is difficult to compare sustainability statements from one geographic location with another. All these factors have slowed down the process and amplified the role of sustainability as a tool of strategic communications (to peers, talent and investors), instead of promoting it as a driver of business growth and resilience planning. That is the impact challenge for the next decade.

Prototyping – as a way to build operational scenarios, validate assumptions and redefine the challenges and opportunities that adopting impact targets entails – can help sustain the learning curve surrounding metrics and outcomes for the long haul.

THE IMPACT CURVE

Let us take for a moment the perspective I bring as an investor entering the 2020s. According to the Global Impact Investing Network (GIIN), the state of

DOI: 10.1201/9781003212225-5   51 the market for institutional *impact assets* – financial commitments for further-ing socio-economic development and sustaining environmental protection – had grown to over US$500 billion in less than 10 years, with nearly 75% of assets lacking short-term liquidity (i.e., they were not easily monetizable). Half the com-mitments were in North America and 20% in Europe. One out of five investors were foundations, as opposed to for-profit organizations. Taking a closer look, the financial return statistics from GIIN highlight that smaller investments (below US$100 million) provide nearly double the economic incentive of larger invest-ments. From an investor's perspective, this seems to imply that the *impact market* is accessible to a limited number of participants, and that it may take 10–15 years before the value of impact, whether financial or societal, can be measured. It also implies that investing large quantities of capital in the impact market may not be appealing to traditional financial investors for long.

But there is more to investment than capital appreciation, and traditional investors are turning their attention to the non-financial returns in the impact spectrum. This will continue to focus their attention on the integration of non-financial sources of data to more clearly articulate a set of acceptable *social and environmental returns*. As with businesses seeking to set sustainability commit-ments, the wealth of public communication surrounding 2030 commitments to the SDGs or to environmental targets is met with early stage reporting of impact outcomes that make the investor's attempt to establish a baseline for quantifiable impact beyond financials even more arduous. In early 2019, the *Harvard Business Review Magazine* highlighted the work of the Bridgespan Group and the Rise Fund in partnership with Y Analytics on a measure known as Impact Multiple of Money (IMM). The IMM employs a social science methodology to provide an estimate of the economic value of targeted societal outcomes, while also evaluat-ing elements of the risks involved in achieving those outcomes. Table 4.1 provides an illustration.

The downside of measures such as the IMM is the number of assumptions and *forward-looking* estimates used to derive the economic value of impact invest-ments to society *today*. Given the relatively small number of established play-ers in the marketplace and their illiquidity characteristics, it is difficult to create a comprehensive narrative about achieving societal and financial benefits. The theory behind measures such as the IMM has created a positive momentum for defining impact outcomes in terms of their size and frequency. This is different from the early focus of supranational entities and international NGOs that drew on progress reports on the basis of the size of the affected/targeted population.

A common saying among traditional investment practitioners is that impact investing as a discipline has been "Thinking Big by Solving Small." That is the main reason why a movement is building up among stakeholders to create a fresh approach to impact evaluation and relevance and to scale the impact faster. From that vantage point, it is clear that impact investors have been doing good, one proj-ect at a time, instead of defining an *impact curve* where socio-economic value is thought of in probabilistic terms – as a function in time subject to the uncertainty of outcomes.

TABLE 4.1

Impact Multiple of Money (IMM)

IMM Calculation Steps	Focus	Tools
I. Assess Relevance and Scale	• Applicable to service/program or process	• Academic journals • Peer-reviewed analyses
II. Identify E&S Target	• Positive and negative externalities • Evidence-based outcomes	• Academic journals • Peer-reviewed analyses
III. Estimate Economic Value to Society	• Comparable sample service/program • Historical field interviews and surveys	• Anchor studies (industry and academia)
IV. Apply the Risk Lens	• Define impact scenarios between research estimates and realization of outcomes • Calibrate estimated economic value by frequency of gaps in early assessment	• Risk mapping and gap analyses • Impact probability modeling
V. Estimate Terminal Value	• Terminal value estimated on the probability that E&S economic added value will continue on a timeframe	• Terminal Value on comparable E&S target
VI. Calculate E&S Return (IMM)	• Value of risk-adjusted E&S economic value-add vs. total investment	• E&S return on comparable product/service/program

Source: Author's intuition adapted from Addy et al. (2019). For illustrative purposes only.
Note: E&S stands for Environmental and Social.

From 2018 to 2020, I experimented with changing the narrative about impact outcomes. I conducted a series of study groups that aimed at reconciling investor and business attitudes toward sustainability commitments and impact evaluation in light of their tolerance for risk – both financial and societal. More often than not, as the dialogue shifted from impact outcomes or sustainability commitments to the risks participants were willing to take that day, the group was more likely to identify the low-hanging fruit (impact opportunities worth pursuing) in the context of each participant's own mental model.

As presented in Figure 4.1, intuitively, the more Environmental (E) and Social (S) investments companies pursue over time, the better the chance of success – the impact curve assumes a more rounded shape. The more that businesses and investors view each opportunity as a discrete, one point in time scenario, the more the impact curve is flattened.

Building from peer reviews and practitioners' feedback on the early hypothesis of the existence of an impact curve, I use the term "curve" to define the trade-offs

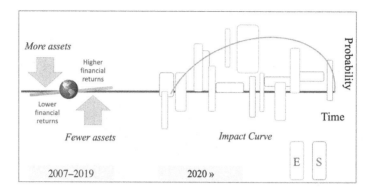

FIGURE 4.1   Building an impact curve. Author's research presented at the Women in Finance Conference and Awards 2020 – Canada, and the proceedings of the 2018 Globe Capital Summit in Toronto.

between economic and societal outcomes that need to balance each other over time with the objective of maximizing both independently. Intuitively, if an opportunity does not yield financial value, it will not sustain societal value either.

The term *impact* is defined as a "change in probability" of the outcome and, therefore, the ability to *impact* the outcome. The value of impact is in changing the probability distribution of an event in time, thus lowering the chance of negative consequences and increasing the chance of positive outcomes over time. This also involves influencing the behavior of consumers, savers, investors and policy makers. In fact, by envisioning impact as a curve that reflects the "risk choices/tolerances" of our society over time, we break "Big Steps" into "Small Steps" and increase the probability of positive outcomes.

By extension, the process of actively incorporating in the impact narrative the map of risk tolerances across ecosystem participants, including those of investors, allows us to better define sustainability commitments, address acceptable time delays and set intermediate goals. It is also inevitably connected with shaping the organizational learning curve of businesses to best integrate impact-oriented goals, aspirations and, ultimately, outcomes in strategic planning. (See Technical Note for mapping risk tolerances to sustainability objectives and impact outcomes.)

PROTOTYPING FOR IMPACT

Impact-centric prototyping equips management teams to commit to the sustainability (E&S) targets that best suit their organization according to the time horizon that meets the needs of their stakeholders. As discussed earlier, developing a framework to prototype with sustainability objectives in mind – to design, test and integrate them in the operating model of a business – aims at weaving the sustainability commitments into the fabric of the organization. It requires identifying, upfront, all interest groups that are directly affected by those commitments – both internal and external stakeholders.

Sustainability targets are increasingly science-based, and there is broad agreement that impact needs to be monitored and measured both within and outside the SDGs. Impact-centric prototyping can help set targets and commitments to address environmental and social externalities that a business is contributing to, either directly or as part of an extensive supply chain network. When applied to the operational and product-specific footprint of a company, impact-centric prototyping activities provide a way to look into the future.

The value of impact-centric prototyping can also be augmented by early and ongoing involvement in engagement activities. This helps the active dialogue between a business and its stakeholders to grow organically and remain representative of the evolving needs of all interest groups. In his tenure as Secretary General of the International Rubber Study Group (IRSG) in Singapore, an international body seeking to improve transparency and strengthen international cooperation in the global rubber industry, Dr. Salvatore Pinizzotto shared his perspective on the role of cooperation in commodity markets such as natural rubber. He believes that when it comes to corporate or government policy decisions, a necessary step is to go beyond the immediate conditions of local plantations, to think about the incentives from public subsidies which do not necessarily teach farmers to be more sustainable or adopt sustainable farming practices. As applicable across agricultural commodities markets, the degree of economic and social development affecting standards of living, gender equity and the safety of women in the workplace and outside are important human capital factors to benchmark "impact" to.

Adding human capital considerations to the assessment of how resilient local agricultural production practices truly are requires a shift in mindset regarding the procurement cycle. A shift from a mindset of sourcing commodities to prototype and meet marketplace demand – in the case of natural rubber, for example, demand in the global automotive industry – to one of an organization that is in "reset mode." From products and services to positioning and market share, an impact-centric prototyping process makes it possible to design a solution that identifies and addresses areas of a business most likely to pose "competing commitments." These are usually areas where financial stakes are at odds with an organization's responsibility for the non-financial externalities it produces. For example, when controversial business activities exist, a prototyping mindset for positive societal externalities would address controversial outcomes first.

The myths of prototyping for impact are worth exploring:

Myth #1 Prototyping is not for every business – it can only be applied to physical products and rarely applies to companies providing "services." I have spent 25 years in financial services and used prototyping without even calling it that.

Myth #2 Prototyping is context-bound and user-specific. Whether it is utilized for new opportunities or to reinvigorate processes and products with a long history, prototyping can be rendered in a context-less setting and thought of as a way to meet the needs of a broad range of users and stakeholders.

The most valuable lesson I have learned in employing prototyping to drive the organizational redesign of processes for sustainability is this: It is a journey that sustains learning at a pace that all stakeholders involved in the exercise can benefit from.

Over the years, the real estate sector has provided insights on how to establish causality between responsible business practices and the creation of socio-economic value. While standards of responsible business practices are well established in the sector – from the management of construction sites to the allocation and continued support of actual properties in the marketplace – assessing the value created and attributing that value to measures of corporate responsibility and social welfare is an evolving area.

In 2010, the Yale Center for the Environment created the first parameters for the evolution of a sustainable real estate index, which is now part of the Dow Jones family of indices. Factors associated with the resilience of a portfolio of real estate assets were correlated with their geographic location as well as their direct impact on climate mitigation efforts over the lifetime of the property. A decade later, the industry started developing a more compelling set of standards. Prototyping began for the impact of social parameters such as tenant wellbeing on the economic benefit associated with a portfolio of properties or a specific real estate site.

Factors such as basic health and safety considerations, the adherence to antiasbestos regulations and other environmental permits, and the role of, for example, commercial lighting solutions on the health of tenants, have become subject to academic studies as well as supporting continued advances in the built environment. Given that the built environment contributes, on average, 80% of total greenhouse gas emissions from urban areas, much attention is given to the efficiency of lighting, insulation and installation of optimized HVAC solutions. Thus, prototyping to address both environmental factors and direct health and community dimensions continues to be key to the development of clear industry standards given the role that urban versus rural areas plays in the socio-economic development of entire countries.

ALIGNING STAKEHOLDER FEEDBACK

Businesses face a common challenge when adopting prototyping techniques to develop impact-aligned sustainability initiatives. That challenge is how to bring a wide range of stakeholders to the prototyping exercise in the right way – a way that focuses on delivering outcomes instead of getting stuck on the organization's existing operational barriers. The following discussion looks at the issue in the context of a highly dispersed supply chain – natural rubber.

A DISCUSSION ON SUPPLY CHAIN DISRUPTIONS DURING COVID-19: THE CASE OF NATURAL RUBBER[1]

During the early days of the COVID-19 crisis, I had an opportunity to study closely the natural rubber supply chain, as documented in a contribution to the

Singapore-based International Rubber Study Group (IRSG) mentioned earlier. While natural rubber production is primarily known for its use in the manufacturing of tires, it is also used for protective equipment utilized in the medical field – an end-market that suffered significantly from a shortage of basic materials during the pandemic. By evaluating the natural rubber production ecosystem, its biodiversity backdrop and the time it takes to go from primary raw material to manufactured good production, it became clear that traditional bottlenecks in the production cycle of tire manufacturers and key weaknesses in promoting environmental and social standards in the procurement process could be solved concurrently. There was a close relationship between these two factors.

The COVID crisis highlighted that close relationship. As emerging economies supplying most of the world's natural rubber products entered lockdowns, responsible harvesting of latex from rubber trees was replaced with fast crop rotations to benefit higher priced agricultural commodities. Automotive manufacturers that chose to address the impact of potential ebbs and flows in their sourcing activities early during the crisis and provided economic incentives to support rubber plantations – whether directly or through industry coalitions – have been able to resume their production cycle more expeditiously than others.

Responsible sourcing can be enacted in an authentic and lasting way where it is proven to be an economically material component of the manufactured good and a key element in maintaining stable supply chains during times of disruption. In effect, while rubber accounts for nearly half of the materials in car tires, synthetic rubber – a plastic polymer – represents the largest component. No wonder that automotive parts businesses have shifted the primary focus of their environmental sustainability programs to the recycling of tires to support their multidecade commitments, with less focus on regional efforts in natural rubber plantations. Adding the materiality lens to prototyping for impact, it may mean rethinking the entry point of the test of responsible practices. Is the most sustainable plantation one that requires pre-existing certifications, or one that is made self-sufficient and economically viable?

When workers' employment and living conditions are part of the environmental goals of businesses, the nature of responsible procurement practices starts to change. The establishment of fair trade standards in agricultural commodity markets becomes a tool to contribute to local development and to support long-term partnerships with farming communities to help stabilize sourcing bottlenecks. It also creates a development model of impact for businesses as opposed to impact-centric or KPI-centric models of environmental and social targets, which are disconnected from the underlying development step.

One of the key features of prototyping for impact in highly fungible commodity markets, such as that of natural rubber, is that *localized dynamics may have global repercussions*. As such, global sustainability goals require local ambition to succeed.

Dr. Pinizzotto stresses the relevance of adopting a localized framework as opposed to choosing a set of global commitments – this can be reflected in the underlying dynamics of the market as well. In the case of natural rubber plantations, commitments historically have focused on the dialogue with farmers and small holders. Nevertheless, economic incentives given to local farmers in an effort by governments to maintain or grow production of the commodity in their region have been far more impactful in driving crop rotation decisions than sustainability commitments or responsible sourcing practices adopted by companies. This suggests that the recent phenomenon of sustainable policies in sourcing commodities may not have the lasting effect they promise toward achieving multidecade corporate commitments. In most cases, business procurement policies surrounding sustainable production require farmers to obtain certifications which remain costly on a local scale and do not necessarily translate to a better market for local producers (e.g., sustainable production is considered a minimum requirement that producers need to meet in order for them to "exist"). Rather, bringing data and basic technology to farming needs and advancing the ability of farmers to increase productivity, crop yields and the like, creates individual incentives for local communities to maintain and grow production responsibly.

THE TROUBLE WITH IMPACT SURVEYS

Impact surveys and impact assessment questionnaires are becoming widespread and increasingly sophisticated across industries. Whether labeled as such or not, I am referring to the bulk of inquiries addressed to organizations of all kinds – corporate entities, government agencies, not-for-profits – which have a governance structure in place and interact with a wide range of stakeholders beyond the direct beneficiaries of the information.

Whether the information is already captured in an organization's impact/sustainability report, on their website or in other standardized format and publicly available, impact surveys and questionnaires usually serve the purpose of the sender. Categorizations, data fields and definitions are not really tailored to suit the organization they are addressed to. That is the real trouble with making sense of inquiries that look for data to substantiate the impact outcomes of a business. This is regardless of the purpose of the inquiry. That inquiry might be to detect trends in business integration of sustainability dimensions, measure the pace of progress towards stated goals, or compare whether impact-oriented activities are building up to scalable results across geographic markets and economic sectors.

When prototyping for impact, project managers may look at the bulk of inquiries and expect the prototyping exercise to drive "alignment" in the feedback that is sought from stakeholders. Let us consider the example of inquiries that point to more transparency (more data, more methodologies, more disclosures) with regard to procurement. In this case, the "prototyping for impact" exercise may be biased towards prioritizing procurement as an area where easy targets can be found – the "low-hanging fruits" within reach and possibly already in the making for businesses. That in itself is a deterrent to a learning enterprise that focuses on impact-centric activities. In fact, more often than not, inquiries may only reflect regulatory pressures and be prompted by the need for regulatory compliance.

A benefit of conducting impact-centric prototyping exercises is that they affect organizational learning. They better align the mental model of all involved in the design of efforts to embed sustainability in making decisions about products in the marketplace. Moreover, they do not require extensive time. The academic article by Gerber and Carroll (2011) provides scientific evidence of the psychological advantage of prototyping in a rapid fashion. It argues that this establishes an environment where failure of an idea in prototype phase is an opportunity for continued learning and refinement of ideas, as opposed to an outright failure that would hinder innovation within the organization.

ITERATING FOR CHANGE

What I have found in my interaction with companies on the sustainability journey and ready to make the jump to "commitment" is that, regardless of their size, their market share, brand recognition or geographic location, more often than not they approach the "exercise" of establishing corporate sustainability commitments by looking at what peers and competitors are doing. While staying abreast of industry advances may provide some direction and highlight a set of opportunities, in the case of sustainability, it can be a quite limiting proposition.

Early movers – those organizations that are bringing transparent and accessible business practices to the implementation of sustainability targets – are setting a baseline for investors, consumers and their own employees. They provide a snapshot of their current standing and thinking on a wide variety of issues – from environmental to community. But there may be few of them and they may not be putting "enough" efforts in place to advance the range of economic activities in their sector to deliver tangible sustainable outcomes. They are examples of localized efforts, of preferences of what to address first. And while they may provide a hint of organizational readiness in specific areas, they lack a sense of urgency.

In many cases, early movers that have been able to push the boundaries of their sector forward have been quick to embrace the idea that a shared discovery – a business practice that is able to endure the test of time and the test of the outcome – is more likely to yield longer term results when addressed as an industry initiative rather than a standalone setting of transparent practices. More often than not, these joint efforts by early movers have required capital investment to support initiatives that span markets and continents and bring together many stakeholders at a table that continues to evolve.

While peer benchmarking may be widely used, it has to be integrated into the discovery phase as a way to identify which businesses to engage with – ones that bring shared values and resources to the table. In a nutshell, benchmarking is likely to discourage businesses that have not yet aligned internal resources to start or support a growing sustainability effort. It can stall internal discussions about best practices and create an environment where sustainability initiatives and efforts are not given the chance to mature and become the testing ground for new business models, new services and products that yield sustainable growth.

When iterating for change, whether the organization is an early mover or latecomer, it is better to start with "aspirational" goals and targets when asking questions and assembling data. Ask questions such as, "What if we aspire to be XYZ, assuming our efforts would not fail?" This is about removing barriers and preventing past conditioning ("we tried but it did not work") and benchmarking ("x number of companies y times larger than us are able to accomplish z by employing a much larger set of resources") from limiting data collection, idea generation and internal feedback. Iterating for change must start with the aspiration, not the constraints or limiting facts about existing capabilities and resources.

INNOVATING AN ECONOMY IN TRANSITION

While start-up incubators and alumni angel investor groups have become a steady feature of many university environments in North America and beyond, home-grown company incubators are a runner up. Sustainability-focused innovation is increasingly attracting pools of capital and talent to work on the next impact challenge: The "transition" economy.

I use the term *transition economy* to refer to companies that need to redefine their business model and the way they deliver products and services in the global marketplace. Specifically, they need to do this because their standard operating processes may create environmental and societal harm.

In the race to zero emissions and closing the gap on social inequities, these businesses must innovate from within and pivot their practices to redefine a holistic value proposition for all stakeholders. Transitioning 80% of global economic activities to their "better self" requires iterating for change. Change in evaluation metrics of business impact, in drivers of stakeholder engagement, and in feedback as direct inputs to the data used to make decisions and align financial incentives to address negative externalities and turn them into contributors of societal value.

When the business lens is reframed from a place of "transition," adopting an impact process centered around iterating for change is key to redefining the scope of current commitments and building the next generation of impactful products and services that responsible consumers will value. Benchmarking to peer groups limits a business's spectrum of influence. But the "transition" lens provides a new whiteboard for businesses that operate in sectors that historically have accepted the negative environmental and social labels as "trade-offs" to provide a much-needed service to society at a specific point in time. When iterating for change, if the desired outcomes are to be achieved, there must be a lens on the tangible need for transition economies as opposed to "trade-offs."

Defining the prototyping exercise with the goal of redefining organizational commitments and business targets in terms of "transition" opens the door for a strategic dialogue focused on new opportunities. Those opportunities can revitalize services, products and business at the sunset into their new value proposition.

Companies and investors are struggling to establish a baseline for quantifiable impact beyond traditional financial measures. While the lack of standardized reporting of environmental and social dimensions has contributed to slow that process down, the art and science of prototyping business scenarios that help visualize how impact outcomes affect their strategic and operational readiness is instrumental to drive the impact movement forward. But there are questions to be answered.

How should businesses start prototyping for impact?

What role does a Chief Financial Officer play in promoting integration of environmental and social outcomes?

THE LEARNING JOURNEY – TAKING IMPACT A STEP FORWARD

The following checklist provides a roadmap to help the reader sketch a preliminary impact curve for an organization of choice and build the foundation of a prototyping exercise.

- Use public statements about the organization's sustainability objectives or sector research on sustainability trends to draft a list of impact outcomes.
- Select environmental and social (E&S) commitments and organizational targets that are best suited to carry out impact-oriented analyses. Research quantitative and qualitative metrics that sustainability reporting organizations such as the SASB/Value Reporting Foundation and the Global Reporting Initiative use for the relevant sector.
- Solicit stakeholder feedback – internal and external – to address operational gaps that may hinder progress toward sustainability objectives. Leverage the network to assess the probability of success given the initiatives currently in place (within the organization or set up by comparable companies).
- Combining stakeholder feedback with the guidance from sustainability reporting frameworks, draft a preliminary impact curve (e.g., list investible areas for product, process and industry partnership opportunities). Iterating for changes in progress toward E&S commitments on the timeline chosen for the impact curve re-addresses the need to set new baseline targets.
- Define how much innovation has played a role in iterating for change. Capture the value of intangibles early and propose metrics to report on the value of ESG intangibles (e.g., use the Yanagi Model from Chapter 2 as a reference to define the value contributed by ESG investments to enterprise value).

TECHNICAL NOTE – MAPPING RISK TOLERANCES TO IMPACT

The first step in building an impact curve of environmental and social opportunities for businesses to tackle and advance their sustainability commitments is to map risk appetite and risk tolerances – the organizational appetite for risk-taking.

This is a key step in engaging with impact objectives that best fit the organization's sustainability journey. As delays between setting objectives and delivering commitments emerge, evaluation of progress and a group's urgency to follow through on commitments is influenced by strategic risk-taking.

INTRODUCTION TO SUSTAINABILITY RISK HEAT MAPS

Risk heat maps are used widely in risk management applications across a variety of fields. A heat map is a visualization tool that guides the user through both a qualitative and quantitative assessment of possible scenarios, ranking risk areas by potential impact and materiality in financial terms. They are increasingly being extended to cover sustainability risks that, while they may not have direct financial consequences on a business in the near term, may pose strategic long-term threats in terms of competitive position or regulatory compliance.

The internal dialogue that occurs with a heat map at hand is often guided by a "residual risk" mindset. Participants discuss broad business resilience and the ability of the existing organizational design and processes to weather the storm if any of the identified risks materialize. Historically, the residual risk lens has been on aligning an effective response to minimize financially material risks and prioritizing efforts based on evaluation of the magnitude and likelihood of events.

When it comes to sustainability risks, many adverse environmental or social factors are best incorporated by employing the materiality framework set by the Sustainability Accounting Standards Board (SASB) and using a sustainability heat map to derive risk indicators in monetary terms as well as assess trends in

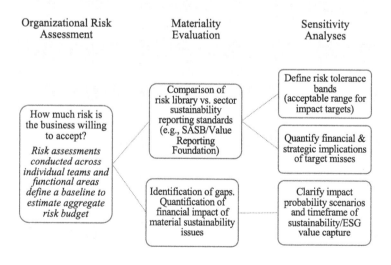

FIGURE 4.2   Phases of the sustainability risk heat map. Adapted from Falsarone (2019).

emerging opportunities surrounding sustainability. When the heat map integrates material sustainability factors, it can also elevate the communication of organizational risks with a set of guidance indicators specifically suited to the sector and the context in which a company operates.

NOTE

1 The example is based on the author's contribution *Horizon of Opportunities – The Investor's View Post COVID-19* to the International Rubber Study Group in Singapore and Vietnam detailing the role of responsible sourcing practices in the global market for natural rubber.

REFERENCES

Addy, C., M. Chorengel, M. Collins, and E. Etzel. 2019. "Calculating the Value of Impact Investing." *Harvard Business Review* (January–February): 102–109. https://hbr.org/2019/01/calculating-the-value-of-impact-investing.

Deininger, M., S. R. Daly, J. C. Lee, C. M. Seifert, and K. H. Sienko. 2019. "Prototyping for Context: Exploring Stakeholder Feedback Based on Prototype Type, Stakeholder Group and Question Type." *Research in Engineering Design* 30: 453–471. doi:10.1007/s00163-019-00317-5.

Falsarone, A. 2019. "Delivering Enterprise Value: The Case of Sustainability and Risk Management Integration." *CIOReview* and *Women in Tech Review*. http://enterprise-risk-management.womenintechreview.com/cxoinsight/delivering-enterprise-value-the-case-for-sustainability-and-risk-management-integration-nid-29273-cid-141.html.

Falsarone, A. 2020. "Horizon of Opportunities – The Investor's View Post COVID-19." *International Rubber Study Group* (April). http://www.rubberstudy.org/content/20671.

Gerber, E., and M. Carroll. 2011. "The Psychological Experience of Prototyping." *Design Studies*. doi:10.1016/j.destud.2011.06.005.

5 The Data-Enabled Sustainability Mindset

To be successful, the sustainability roadmap adopted by businesses needs to be powered by data. So far, we have laid out the journey that organizations seeking to integrate impact outcomes as part of their DNA – and, ultimately, their enterprise value – need to embark on along with their stakeholders (the producers and recipients of societal and financial externalities). A sustainability-oriented mindset is increasingly perceived as the driver behind a global economic transition. When data-enabled, it builds on the value of transparent decision-making and actionable insights to deliver impact targets.

How have the businesses that made an early start on this journey managed to effectively incorporate impact-oriented data and frameworks in their decision-making? Have they prioritized governance policies that align with their operating environment, and with their customer, employee and community relationships, to balance financial and societal trade-offs? How do they enable continuous learning? What are the metrics they live and breathe by? How have they aligned their risk culture with organizational incentives and accountability that drive continued learning and a dynamic approach to transforming their business model to be truly sustainable – ready to withstand the risks and capture the opportunities of the twenty-first century?

The organizations in the fast lane seem to have a common thread: They drive change from within by building a culture of accountability and risk-taking aligned with their organizational purpose and mission. In doing so, their leadership teams have created an environment where data-enabled decisions can drive alignment between short-term *responses* and long-term *objectives*: From operations, to finance, to corporate development and talent retention. It is humbling to see how these leaders not only continue to advocate for change from within but also steadily dismiss industry pressures and short-termism to build enterprise value *sustainably*.

In researching the impact challenge of businesses, I have been profoundly affected to discover that most of the ones I reference are facing significant hurdles. Hurdles in the geographic markets in which they operate and in the over-regulated sectors they navigate. Hurdles in the form of continued pressure from investors and civil society organizations to build faster and cheaper solutions in an inclusive manner and create deeper outreach and expand their services and products to underdeveloped communities.

I will review the inclusivity paradigm of sustainable businesses in Chapter 7, but it is important to note that the early efforts of a company to gather, analyze, verify and sustain the integration of decision-useful data and insights reflect how

DOI: 10.1201/9781003212225-6

inclusive the stakeholder networks are that are involved in that effort. It is also foundational to building an organizational culture and incentives plan around sustainability targets and corporate commitments that rely on individual actions and group decisions to deliver value in financial and societal terms. Therefore, it will not come as a surprise that the organizations featured in this chapter have also sought and built a highly diverse network of advisors, thought partners and community leaders to help their teams deliver value by iterating for change and embracing innovation responsibly.

THE CFO LENS: THE EXPERIENCE OF ENEL AMERICAS

In early 2020, I had the pleasure of engaging with Dr. Aurelio Bustilho de Oliveira, Chief Financial Officer of Enel Americas S.A., the largest privately owned energy company in Latin America and subsidiary of one of the world's leading integrated electricity and gas operators, Enel Group. I was inspired by the value his team placed on corporate transparency and the ability of management to keep the focus on the local context while also fulfilling the expectations and ambitions of sustained growth and global recognition of their parent entity.

The interview Q&A I conducted with Dr. Bustilho highlights the evolution of a journey from accounting to business innovation that underpins the intentionality of building organizational awareness through data, introducing novel metrics and eventually leading the transition economy through a culture of partnerships and innovation. I engaged Dr. Bustilho in a discussion of ten key areas:

SUSTAINABILITY IN DECISION-MAKING

Q. Where has data played the biggest role in embedding sustainability in your decision-making? Could you share an example where looking beyond short-term business trade-offs has led to the creation of sustained value – even if intangible in accounting terms – either through partnerships, greater adoption of service in local communities or in geographies where supportive regulation has been lagging compared to business innovation?

A. In the previous years, most of our investments were destined for the grids to be more resilient, more efficient, more digitalized and automated in our operations. With the integration of the renewable energy generation assets of Enel Green Power (EGP) Americas into Enel Americas' perimeter, we will consolidate ourselves as a fully integrated and sustainable business model.

Obviously, there is always room for improvements in our traditional business. Now, even more, you can integrate the traditional business with the NCRE [Non-Conventional Renewable Energies], giving special attention to circular economy opportunities.

For example, data plays a key role in getting to know your customers. In Brazil, we studied our customers' payments behavior, because it is not enough to sell electricity to your customers. You need to know if they can pay their bills and in what circumstance they are after paying their bills (it makes no sense for them to pay the electricity bill and be left without money for anything else). So we implemented a circular economy project in zones of more vulnerability, in which the customer could bring plastic that they collect from their residuals or garbage. We weigh this and pay the amount in their electricity bill. So in summary, we are applying a discount on their bills, we sell the plastic for recycling, and this is a win-win for the customers, the environment and the company.

INCLUSIVE MINDSET AND ORGANIZATIONAL BUY-IN

Q. How have you helped the regional organization follow the sustainability commitments of Enel Americas by adopting an inclusive mindset and getting the buy-in from within?

A. The CFO must play a key role in the organization to help to understand the importance of sustainability in the organization. Sustainability in the next 10 years will be the way to do business, and it's clear that finance is a service on the way to do business. So this matching needs to be found and enlarged as we are at the beginning of this revolution, this transition. And we must do everything on all the instruments that we will develop.

The Group chose this new way to do business 5 years ago, so it was decided not to have some sustainable projects, but instead put sustainability as the company's purpose to create value along this path. Thus, we analyze sustainable finance with an innovative approach to link the company's purpose with a sustainable finance instrument. That's why Enel created the SDG-linked bond that is a different way to do sustainable finance. It is linked to targeting and not linked to a specific project, because we think that sustainable finance must have a holistic approach.

THE ROLE OF DATA

Q. Has data proven to be a value-add in your discussions with regulators and investors? What kind of data has had the most value in supporting internally your sustainability commitments?

A. Like all the new products or services, the first thing you must do is a pilot, then analyze the data collected and the results; this information is how we implement the smart meters.

For example, the regulators must understand why it is essential to include the smart meters into the RAB [Regulatory Asset Base[1]]; you must give them accurate data and explain why it is important to the end-user to have them. As mentioned, the rollout program in Sao Paulo started with a pilot. Once the pilot was finished, the regulator allowed us to do the massive rollout because they understood that the company is not the only one that benefits. The customers also benefit.

TRUST ENVIRONMENT

Q. Have you created working groups and built a culture for understanding sustainability from the ground up – involving regional managers, electric utility commissions or even peers in engaging in a broad dialogue supportive of green energy?

A. As I mentioned before, sustainability is the company's purpose. Five years ago, Enel adopted this new way to do business, and we, as part of Enel Group, pursue this purpose. And with the integration of Enel Green Power, we will provide help and solutions to the countries in which we operate regarding the benefits of the energy transition and the need to decarbonize electricity; we need to tackle the Paris Agreement. On this, we are all together; no country can be left behind.

STAKEHOLDER DIALOGUE

Q. How have you dealt with comparing legacy businesses versus your vision of the future, when legacy also means job security, known revenue streams and long-held regional relationships with customers?

A. Yes, you know that Enel Americas was created in 2016 when the Group decided to isolate the Chilean business from the rest of Latin America because it was a more mature market than other countries.
   So that is our legacy business: distribution, and conventional generation; now, from the first of April, we are consolidating the NCRE business from Enel Green Power Americas. This "non-conventional" business now is more "conventional" than ever; it is the future of the generation business. This operation will not affect our customers' relations because EGP Americas was fully contracted with Enel Americas.

INTEGRATION OF NEW DATA SOURCES

Q. Can you share the example of a data exercise or the mapping of data that has shed light on the ability of your business to honor its legacy while moving into the future? Have you been able to use "facts and figures" type of reasoning at the decision table with advisors, boards, customers and regulators to build support for your decisions? Any new data points in your management dashboards that you believe are instrumental for a CFO to drive the sustainability agenda of a company in your sector in a way that it is also balancing the societal and financial trade-offs that you could not live without?

A. I will say the CO_2 emissions; this is the past and future in all senses. Countries and companies (public and private) must be pointing to the net zero emissions commitments by 2050 or even before that year. The CO_2 emissions come from our legacy; you know that thermal generation is the only one that emits CO_2. The CO_2 is something that the group uses in sustainable finance; you know that Enel linked a bond issuance to a target of CO_2 by 2030; this is the given example of the sustainability that came into the perimeter of the CFO.

Learning Journey

Q. What is the biggest lesson(s) you are learning in the process?

A. A company without a long-term strategy will be extinguished; you need to take the train urgently to the net zero commitment. Otherwise, you will disappear; maybe we don't know the path to reach the target, but you will discover it on the way.

Strategic Communication

Q. How have you been able to bring along the most difficult stakeholders in your quest to deliver sustainable value creation? Has transparency and intentionality in laying out your strategy, communicating your rationale for change and recognizing any "mistakes" early helped in transitioning your organization to embrace such an impactful business model?

A. We think that communication and transparency are two of the most important things regarding a healthier relationship with your stakeholders. You must involve the local community before you make an investment decision; you need to listen to their needs, and what they expect from your project. That's why we use the CSV [Creating Shared Value] approach in our projects. Before this we used the conventional approach – approving a project without involving the local community; there was a financial valuation only.

Data-Driven Decisions

Q. What do you see as the power of data insights in continuing to support your business and investment case to honor your sustainability commitments and make them central to your DNA?

A. In the new world, without data, you are nothing. You need to know your clients; you cannot treat them like they all are the same.

This is what we are pursuing. For example, with the implementation of smart meters, you can know your customer's needs and not sell them just commodities.

TRANSPARENCY AND OPEN DATA

Q. In a world that pushes for open data architecture and the ability for all to access data that supports business decisions, how do you feel about open data for your sector where regulation has played such a distinctive role in opening pockets of opportunities between regulated and deregulated markets (specifically in the Latin American context)?

A. As I mentioned before, the data will play a key role in the new world; we are open to sharing but with the customers' consent because, in the end, it's their behavior that you are sharing with the market. That's why the regulators in Latin America must improve the data protection system to be at the same level as the mature markets.

Among other things, Dr. Bustilho highlights the value of embedding sustainability as an organizational value into the enterprise value of a company; the need for boards to be educated on ESG topics and enhance governance to reflect sustainability risks and opportunities in the ordinary course of doing business. It underscores the urgency of building dashboards to inform decisions on clearly identified sustainability metrics that drive corporate strategy. It also demonstrates the importance of sharing data in a trust environment and driving an organizational mindset focused on delivering sustainability returns.

LEVERAGING SUSTAINABILITY DATA IN STRATEGIC PLANNING

Over the past 15 years, sustainability risks have ignited the dialogue between management teams, boards of directors and special interest groups. From environmental catastrophes associated with oil drilling to the myriad recalls affecting auto manufacturers, governance responsibilities in the context of E&S principles have become directly correlated with upfront costs rather than value creation.

Are boards ready to take into consideration the next set of sustainability factors? Strategic scenario planning in the context of sustainability matters requires boards and management teams to think in terms of foregone opportunities, not outright costs and the net effect on topline growth if a company chooses *not* to engage. For example: How would new business wins be affected if a company does not commit to environmental safety investments? Focusing on the remote likelihood of an event inherently justifies today's widespread inaction. Scenario thinking changes perspectives and helps redefine the baseline (Table 5.1).

The call to make sustainability a fiduciary duty of leadership teams and their overseers – board members – has become stronger and stronger over the past decade. The Sustainability Accounting Standards Board (SASB) has pushed

TABLE 5.1
Planning Approaches and the Integration of Sustainability Objectives

Characteristics:	Traditional Planning	Scenario Planning	Sustainability Integration
Mindset	"Everything else being equal"	"What if everything else is different?"	Science-based, time-bound, collaborative
Metrics	Mostly quantitative, known, objective	Qualitative/ quantitative, subjective, may be unknown	Non-traditional, science-based, time-bound
Relationships	Stable, established, numerical	Dynamic, open to feedback and to new entrants	Cross sectors/cross disciplines, innovative
Approach to future	Adaptive, passive (future events will occur)	Multiple, uncertain outcomes (future events can be analyzed)	Compelling, active, urgent (future events influenced by today's decisions)

Source: Author's own intuition. Author also referenced Lindgren and Bandhold (2002). For illustrative purpose only.

to increase companies' voluntary disclosure of sustainability metrics to adopt materiality criteria in the selection of both qualitative and quantitative assessments. Thanks to these efforts, governments, stock exchanges and others are following suit and moving to identify mandatory reporting standards for companies. Depending on the economic sector in which a business operates, boards are increasingly addressing E&S commitments through the lens of materiality assessments. These are added as an annual review of the risks and opportunities affecting the business directly or through its ecosystem.

By the end of 2020, over 400 early-stage businesses had come to market to offer some form of sustainability analytics. These included, in some cases, verification and attestation services to support the validity of the underlying analytics, most of which rely on a mix of artificial intelligence, big data and cloud services. Many organizations have commented on the lack of transparency and depth of some offerings, while others have pointed to the frustration of dealing with third-party sustainability ratings or ESG scores.

It is no surprise that we have started the next decade with much more data at hand but fewer reliable insights. What is the procurement team of a company to do when seeking to provide transparency and add impact-oriented data to their efforts? It is understandable how some companies have chosen to disclose less and focus on the quality of disclosures instead of inundating the marketplace with additional data points. While the world of sustainability data is likely to increase in complexity, it may continue to stay "time sensitive" given the potential latency of public data reporting.

For the purpose of this discussion, I refer to the challenge posed by data "latency" as *the time elapsed between when observations are collected and when they are available for analysis,* including verification and reporting. There are stark differences among metrics derived from the sustainability data universe. Some are derived from environmental factors; some have well-known unit measures and common terminology; others are emerging social impact metrics involving data privacy and employee wellbeing. Given this, latency is an important characteristic of the data that many users – from corporate decision makers to the general public – face in drawing causality insights or compiling sustainability dashboards. The risk of applying too narrow a definition to "real-time data" is that of low-latency of metrics. In this case, quicker operational routines to flag early warning events may ultimately lack quality and scientific support. (See Technical Note for data latency applications in climate assessments.)

THE ROLE OF INTEGRATED REPORTING[2]

The year 2020 was truly the year of data for E&S Governance (ESG). The entire finance and management accounting fields – from corporate finance specialists, controllers, treasurers and CFOs to auditors and investors – took a closer look at sustainability data generated across their businesses. They started to take stock of any causality between environmental or social credentials that a company may be awarded among its industry peers, and the sustainability factors that affect the profitability of their business. Where sustainability can add the most value in reporting is at the intersection of enterprise value, operational efficiency and top line growth. According to PwC's 2020 corporate directors survey, more than half of corporate directors said investors were giving too much time and focus to environmental and social governance considerations – which is nearly twice the percentage identified in the same survey two years earlier. When we compare that survey to the engagement priorities for institutional investors during the 2020 proxy season, the focus remains on environmental resilience and also the ability of the board to address capital planning accordingly. Yet, few enlightened board chairs or audit committee members seem to be keeping the accounting for ESG data high on the priority list.

Concurrently with the release of the board surveys and the 2020 proxy season results, I had the opportunity to share my perspective on the future of integrated reporting and ESG data with the Institute of Management Accountants. The IMA® asked me a series of questions in a podcast conducted in April 2020.

Q. What are some of the challenges that accountants face with ESG data when it comes to giving investors the information that they're really looking for?

A. When I think about financial innovation and the headlines surrounding the issuance of sustainability-linked financial instruments such as green bonds, loans or even transition-linked capital raising instruments, they have certainly raised awareness as the treasury team at a company and the CFO is as involved as the auditors and the external verifiers in aligning capital raised with the investments associated with a company's environmental efforts. Certainly, from an accounting perspective, the need for ESG data is to be aligned with the financial commitments of a firm, as data is about long-term trends that will require as much operational expenses as capital expenses to build resilience or competitive advantage and top line growth. Either way, ESG data and the impact on dollar unit measures are the hardest to address, as non-financial risks and non-traditional sources of risk don't come in the same unit measure.

Clearly you have accounting and finance professionals testing themselves on kilowatt-hours when aggregating energy efficiency to metric tons of CO_2 per home yearly when discussing home energy use, or even more esoteric measures such as "near misses," which is the count of events with the potential of loss or injury if the accountant is analyzing health and safety statistics within the workforce. This is increasingly the domain of financial professionals. While there is certainly room to define best practices, it is simply good business management to define ESG indicators at the company level that are associated with financial outcomes and address them and report them consistently – they could be in the form of trends or as static, absolute levels if there are (sustainability) targets in place.

A recent example is that of a Colombian green cement laboratory, which calls for the relevance of adopting green techniques in the production of cement while production itself is clearly an energy-intensive process. So while the green aspect of bringing to market a more sustainable product in terms of carbon released in the atmosphere during manufacturing may indeed be part of an environmental materiality approach and disclosure that can be externally verified – with clear financial consequences associated with potential carbon levies on that economic activity in the future – the financially relevant aspects to be accounted for are associated with other issues.

First, business materiality would call for near-term financing of that transition from legacy cement production to newer products, including timing of operating and capital expenses. Second, the forecast of market share gains and therefore more stable pick-up in free cash flows including the impact on credit metrics.

Unless we place the financial materiality of the innovation in the context of how sizeable is the opportunity, the financial benefit associated with greening products may not be fully captured in enterprise value and therefore there may be no additional incentives for a management team to carry out a full upgrade of their product offerings.

The accounting world has the remarkable opportunity to pick up on innovations and articulate that in terms of financial outcome, with a reporting time horizon in mind. Moving from sustainability credentials, awards and recognitions into a set of variables that are really part of the everyday investor dialogue is key. Certainly, the underwriting of green bonds and sustainability-linked loans and the advances in capital markets addressing the quality of project financing with green aspirations has made corporate commitments even more visible globally.

Q. What is the difference between integrated reporting compared to the traditional financial reporting that most accountants are familiar with?

A. The value of integrated reporting is more tangible than ever when dealing with sustainability factors that affect a business in ways that are certainly not uni-dimensional. Given the different unit measures between financial and non-financial variables and the continued need for comparability, the value of having a solid corporate reporting that incorporates the governance of environmental and social factors in management discussion and analysis is key. In addition, as reference to E&S metrics gets incorporated in financial commentaries where material, a well-structured integrated reporting process can shield a business from the risk of miscommunication or lack of targeted communication surrounding financially relevant issues of strategic value to an organization. According to a recent study by the Alliance for Corporate Transparency, a majority of companies are focused on policies and procedures and commitments with respect to non-financial risks such as sustainability, but only one in four report on them and define performance indicators that are verifiable.

If we delve into issues related to the workforce, the statistics are even more alarming. Approximately 80–83% describe their human rights policies, 25% disclose the actual risks related to human capital and less than 15% report actual impacts on the business. In the Ernst & Young report on investor engagement priorities for 2020, human capital retains the top two positions along with environmental concerns as a key strategic focus area in the next 3–5-year time horizon, including issues of workforce diversity and culture in the workplace. Yet, traditionally, investors have focused on workforce compensation in their analyses, as in many cases that opens the dialogue on pay equity and diversity in promotion rates. While not a one-size-fits-all approach, it is highly dependent on the end markets in which a company operates, and reporting should clearly be aligned with the need for more accurate statistics.[3]

Q. How has technology impacted integrated reporting, data accuracy and comparability?

A. Technology has been a driving force of business transformation in the business management and services industries for the past two decades. Going forward, technological innovation is likely to contribute to the sustainability journey of organizations in two ways. First, through enhanced verification and assurance. Second, directly through the incorporation of non-traditional sources of sustainability-oriented data feeds into financial analyses and strategic planning.

With respect to the first application, there are a vast number of data points with consistent history and verifiable sources of information available today. This has created the best backdrop and a strong tailwind to harness use cases of alternative data and other non-traditionally collected nor accounted-for data series that, in aggregate, enable decision-useful content in the reporting of business outcomes.

An example that comes to mind is the emerging use of geospatial data or earth observation through remote sensing technology for the analysis of environmental risks in supply chains. It is clear that such applications will become core in financial reporting that is related, for example, to enterprise risk management and scenario planning under the Task Force for Climate-Related Financial Disclosures (TCFD).

That brings us to the second leverage point of technology innovation in accounting and reporting applications – the translation of non-financial data into financial metrics.[4] The real need here is to ensure that whatever technology provider is selected or in-house platform is rolled out, it is conducted from a place that truly looks at the ultimate use of data. There are monetary liabilities associated with mismanagement of information. How much of that information has a verifiable process to derive financially material decisions for a business? It may also be subject to cyber and broader reputational risks, even if it comes in non-financial unit measures and looks harmless. If we think of medical records and privacy, it is easier to see the case for heightened vendor management practices surrounding the use of technology for sustainability information.

An area where technology will become increasingly relevant is the third-party verification of impact. This is yet to be addressed and likely a crucial aspect in a world where companies set bold climate science-aligned targets for carbon emissions and start discussing variables such as Scope 3 emissions. Such variables are not measurable as part of operational frameworks but go outside of the company (the use phase) and follow the products or services once they leave the showroom or the shelf.

So far, the collective work of business leaders and investors has contributed to deploy a series of frameworks, metrics and process enhancements that help organizations jump start their impact journey. A more programmatic approach to set impact objectives and deliver sustainability returns along with business profits requires some deeper introspection.

How does organizational culture affect the impact journey?

How can we turn the impact challenge into an opportunity for lasting adoption of sustainability innovation beyond metrics and rankings?

THE LEARNING JOURNEY – THE DATA-ENABLED SUSTAINABILITY MINDSET

Impact outcomes are business outcomes. By working step-by-step through the following checklist, I encourage the reader to unveil the value of sustainability data in driving business insights and promoting integrated reporting of E&S information along with traditional financial metrics.

*   Define the primary data needs and the analytical insights to adopt a sustainability framework that touches every aspect of the near-term business strategy and its longer-term E&S commitments.
*   Visualize data needs by addressing gaps in data-driven business insights of sustainability metrics.
*   Address any variability in time-dependent metrics and evaluate latency-driven gaps.
*   Set up an accountability-oriented roadmap to test intermediate-term insights derived from data at hand.
*   Take stock of operational successes and failures across products, regions and customer segments and learning behind perceived failure.
*   Set guideposts/reminders that E&S commitments require an organizational alignment built on translating the E&S commitments into internal roadmaps across functions (they do not only reside with the Chief Procurement Officer or the Chief Sustainability Officer). Accountability of leadership and board members is a key driver of success.

TECHNICAL NOTE – DATA LATENCY IN CLIMATE ASSESSMENTS

The introduction of a framework for climate-related financial disclosures by businesses (also known as TCFD) in 2015 and the urgency of aligning consistent measurement and management of climate risks has been followed by significant efforts by practitioners and academics on how to integrate quantitative data assessment of climate events in the more traditional business context. One particular overarching question has been about the acceptable degree of latency when integrating climate analysis data to assess environmental sustainability of both private and public sector participants (Table 5.2).

TABLE 5.2
Data Latency and TCFD Reporting

Latency Categories	Degree of Acceptability of Climate Data Latency for TCFD Reporting Alignment			
Minimal data latency (e.g., real-time) *Near-time* data latency (e.g., at set time intervals)	Governance: Disclosure of governance practices related to the company's climate risks and opportunities	Strategy: Integration of potential impacts of climate risks and opportunities on the business (materiality)	Risk Management: How the business identifies, evaluates and manages climate risks	Metrics & Targets: Disclosure of the frameworks used to manage material climate events
Some-time data latency (e.g., updated as needed)	*Degree of acceptability:* **Near-time data latency** *or better is adequate*	*Degree of acceptability:* **Near-time data latency** *or better is adequate*	*Degree of acceptability:* **Minimal data latency** *is required*	*Degree of acceptability:* **Minimal data latency** *is required*

Source: Author's intuition. Data latency categories adapted from Bison Analytics.

The topic of data latency has gained even more traction as companies have pledged their commitment to carbon neutral pathways in their operational and product footprint, which includes their procurement decisions. As government commitments to "net zero" economies by 2050 continue, the balance between greenhouse gas emissions produced and those removed from the atmosphere has shifted the attention from discrete precision on climate data utilized to a dynamic assessment of progress over 20–30 years. As estimates of climate sensitivity of businesses enter their strategic scenario planning in alignment with the recommendations set by the TCFD, evaluating the *spectrum* of data latency in climate analyses adds to corporate directionality and the overall monitoring of progress.

BALANCING TRADE-OFFS WITH GREEN COMPUTING

While the degree of alignment of climate data analyses with the recommendations set by the TCFD varies depending on the complexity of business operations, it becomes particularly relevant when addressing procurement decisions surrounding renewable energy sources for energy-intensive sectors.

In the proceedings of the ACM SIGCOMM 2012 conference on applications, technologies, architectures and protocols for computer communication, researchers from the University of Waterloo, Canada, investigated green computing in the context of content distribution centers such as geographically distributed servers and their consumption of electricity – for both powering and cooling functions. While data centers require very low data latency for accessibility, they are increasingly under pressure to balance the energy usage of their servers and that of their networks.

Their FORTE framework (Flow Optimization-based network for Request-routing and Traffic Engineering) provides an example of a principled approach to the trade-offs posed by access latency, electricity costs and carbon footprint. One of the novel elements the method introduces is the use of the cost of carbon emission reduction for large-scale Internet services. In fact, given existing differences by location and time-zone of carbon emission rates, businesses that rely on Internet providers across multiple datacenters can successfully reduce the carbon footprint of their service by choosing to redirect traffic to "greener" locations.[5]

NOTES

1   The RAB refers to a long-term tariff system aimed primarily at encouraging capital toward the modernization of infrastructure. It is widely used in the electric utilities sector.
2   Adapted from the author's interview hosted by the Institute of Management Accountants (IMA®) Count Me In® podcast on March 16, 2020. https://podcast.imanet.org/55.
3   The research project on Human Capital that the Sustainability Accounting Standards Board (SASB) launched in the fourth quarter of 2019 provides a reference for cross-sector assessments. It addresses the need for a deeper exploration of dimensions such as human wellbeing in the workplace and the impact of technology and innovation in upskilling, as well as capturing the impact of D&I initiatives on financial materiality.
4   In Chapter 2 we discussed the sustainability journey of Eisai Co. Ltd. Their partnership with business transformation firm ABeam Consulting to scale up human capital metrics relied on a sophisticated visualization dashboard to aggregate data and insights across business segments and global subsidiaries.
5   For a review of the FORTE methodology for businesses, P. Gao et al. present the impact on its application to Akamai, the world's largest content delivery network.

REFERENCES

Alliance for Corporate Transparency. 2019. "2019 Report: Analysis of Sustainability Reports of 1,000 Companies Pursuant to the EU Non-Financial Reporting Directive." https://www.allianceforcorporatetransparency.org/assets/2019_Research_Report%20_Alliance_for_Corporate_Transparency-7d9802a0c18c9f13017d686481 bd2d6c6886fea6d9e9c7a5c3cfafea8a48b1c7.pdf.
Falsarone, A. 2017. "Discounting Materiality for a Sustainable Future." *NACD Directorship Magazine* (May/June).
Gao, X. P., A. R. Curtis, B. Wong, and S. Keshav. 2012. "It's Not Easy Being Green." *SIGCOM* 12 (August): 211–222. doi:10.1145/2342356.2342398.
Google Cloud. 2021. "Sharing Carbon-Free Energy Percentage for Google Cloud Regions." 17 March 2021. https://cloud.google.com/blog/topics/sustainability/sharing-carbon-free-energy-percentage-for-google-cloud-regions.
Lindgren, M and H. Bandhold. 2002. *Scenario Planning: The Link Between Future and Strategy*. UK: Palgrave Macmillan.
National Aeronautics and Space Administration, Earth Observing System Data and Information System. 2020. "What is Data Latency?" 28 August 2020. https://earthdata.nasa.gov/learn/backgrounders/data-latency.

Platnick, S. 2017. "Summary of the Workshop on Time-Sensitive Applications of NASA Data." *The Earth Observer* 29 (2). https://eospso.nasa.gov/sites/default/files/eo_pdfs/March%20April%202017%20color%20508.pdf.

PwC. 2019. "Annual Corporate Directors Survey." https://www.pwc.com/us/en/services/governance-insights-center/library/annual-corporate-directors-survey.html.

Smith, J. 2020. "What Investors Expect from the 2020 Proxy Season." *EY*, 5 February 2020. https://www.ey.com/en_us/board-matters/what-investors-expect-from-the-2020-proxy-season.

6 Culture and Climate Change

Environmental sustainability is an evolving field. What used to be the realm of environmental science majors is now a field at the intersection of business, finance, government policy, economics, engineering, biology and chemistry, to name just a few. The list can go on and on if we delve into the impact of environmental sustainability in daily economic and social activities and whether we are prioritizing one over the other. It is also the field that has seen the most multistakeholder initiatives regarding the role of businesses and civil society in addressing climate change.

With headlines remarking on the emergency of the issue, environmental commitments and carbon reduction targets have been elevated to the agenda of boardrooms and make up a good part of corporate communications and investor relations. Yet in most cases, the attitude continues to be defensive instead of expressing a widespread and genuine desire to reset business and societal priorities. The COVID crisis of 2020 shifted attention to how organizations react to catastrophe planning and resilience building in the very near term, and this may have created a more supportive backdrop against which businesses can tackle environmental disasters.

CULTURE AND SEMANTICS: DEFINITIONS OF CLIMATE CHANGE

In his book *Don't Even Think About It: Why Our Brains are Wired to Ignore Climate Change*, George Marshall, a recognized expert in climate change communications, argues that humankind may have numbed itself to the fear of a climate tragedy. That it may be in denial about the scientific evidence and the consequences of climate change on daily living. The psychology of climate change as an impossible problem is far from being fully comprehensible to any of us. Nevertheless, the dearth of common semantics and definitions has made the presentation of environmental impacts, rather than the science itself, nearly impenetrable. When businesses lack adequate tools and frameworks to discuss environmental impact strategically, minimal disclosure and minimal communication are to be expected – the type of thing that is best expressed through boilerplate statements and legal disclaimers.

When looked at through that lens, the current movement to define common grounds in sustainability metrics and standards and to prioritize a set of definitions is aligned with Marshall's thinking. The Sustainability Accounting Standards Board (SASB), the international body that has pioneered the

DOI: 10.1201/9781003212225-7

incorporation of sustainability metrics in corporate reporting practices, evolved into the Value Reporting Foundation in November 2020. The Foundation is set to redefine how management accountability and corporate commitments go hand in hand with transparent handling of climate risks and opportunities. The creation of the Value Reporting Foundation marks the shift from the rigor of accounting to the value of transparency that drives enterprise value and business accountability.

When businesses choose to mitigate and adapt to environmental scenarios, it is vital that they rely on a common language to reframe near-term and long-term impacts on organizational resilience, risks and opportunities. I argue that the needs of businesses must be translated to an awareness of how environmental events are likely to affect the workforce, the customers and a company's very own ecosystem. And this must be done in a way that is easy to grasp for everyone involved.

Physical climate-related events and environmental disasters are the ones that most often come to mind. But as we move from insurance planning for catastrophic events to the sphere of more frequent and steady changes in environmental conditions that populations around the world are witnessing, we also need to consider the emerging risks associated with the transition of living and working conditions that these changes create, in both economic and social terms.

Building a culture around climate resilience means addressing both direct and indirect impacts of those subtle changes and clearly articulating the scenarios that a business is considering when placing climate change as an organizational priority. In fact:

When an organization discusses water consumption, preservation and wastewater, is it discussing the impact of climate change on its operations?

When it discusses food security and pricing of commodity markets in its procurement cycle, is it addressing climate change adaptation of its suppliers?

When a municipality is struggling with the cost of waste collection and its waste disposal cycle, is it integrating the need for environmental transition in its urban planning activities?

More often than not, the answer to questions like these is yes. Institutions that are focusing time and effort on the consumption, recycling and renewal of the natural resources they have at their disposal are in fact building awareness of direct environmental impacts of changing climate patterns on their operations, workforce, customers and supply chain.

It is remarkable how, by directly focusing on the disruptions caused by climate change rather than on differences in semantics, beliefs and definitions, an organization can build a culture of near-term solutions and long-term resilience while also developing an immunity for paralyzing denials or political statements that have been historically associated with climate risks. Definitions, problem reframing and semantics evolve with time and as scientific findings point to new alternatives or develop low-cost versions of climate innovations.

When a common language for sustainability fails to evolve and to reflect the local context, the opposite may happen. One example is the use of terms such as "forest," "deforestation" and "forest degradation" in the context of international climate agreements and the adoption of sustainable forest management practices. In their contribution to the Society for Conservation Biology, Prof. Nophea Sasaki (Asian Institute of Technology/Harvard University) and Prof. Francis E. Putz (University of Florida) highlight the repercussions of adopting a one-size-fits-all approach to definitions of the environmental ecosystem.

The example they bring is the failure to distinguish between "natural forests" and "plantations" in international standards such as the United Nations Framework Convention on Climate Change (UNFCCC). This has resulted in restrictive canopy requirements that define a "forest," thus leaving the impact of forest degradation unaccounted for and putting basic ecosystem functions for forestry production and biodiversity preservation at risk.

Even when the connection between climate impact and culture is understood well, there can be a lack of urgency when it comes to taking action. This reflects a lag between cognitive acceptance and mobilization of efforts by businesses. A good example of the interconnection between culture and climate change awareness is the body of research carried out in Japan that addresses the impact of changing weather patterns on the timing of cherry blossoms. The Cherry Blossom Festival is celebrated as an important tradition and is part of the country's cultural heritage and attitudes. But behaviors around the connection between earlier timing of flowering as a result of global warming and planning of festival activities in a business-as-usual mode suggest that stakeholders are not taking it seriously. In fact, only organizations whose income is dependent on the festivals are eager to embrace an earlier schedule than tradition would allow. Businesses and local communities are standing at a crossroad in terms of preventive actions and the adaptation needed to adjust to the tangible deterioration of both horticultural and economic backdrops.

NACHHALTIGKEIT: ENVIRONMENTAL SUSTAINABILITY FOR BUSINESSES

Now that we have become familiar with the role that organizational culture can play as a driver of environmental impact for businesses, we can deploy it as the primary engine to reframe the role of environmental action in decision-making.

Not coincidentally, the evolution of modern environmental sustainability traces back to the publication of an eighteenth-century handbook in Germany on the science of forestry management by Hans Carl von Carlowitz. The book addressed the issue of harvesting in a responsible manner to generate a "sustained yield." *Nachhaltigkeit* – a German synonym for the English word "sustainability" – expresses the tension between the health and development of our planet's population and that of its biosphere. It took the following two centuries to embrace a broader definition that would encompass all biological systems (not only forest land) and for environmental sustainability to re-emerge as balancing the ongoing needs of populations with issues affecting the health of ecosystems and the dangers posed by environmental degradation.

In 2020, the EU Taxonomy, the body of work authored by the European Commission to address its 2030 climate and energy targets, became the first attempt to define and categorize "green activities" for businesses. The Commission has introduced a roadmap for economic development aligned with a series of environmental commitments which include, among others, climate change mitigation and adaptation, the role of the circular economy, control and prevention of pollution, and protection of biodiversity (Figure 6.1).

In its very first year of implementation, the taxonomy has given life to numerous multistakeholder regulatory actions and policy decisions. These initiatives are designed to guide businesses in their transition from their current environmental and societal footprint to one with a low carbon ecosystem that drastically reduces carbon emissions year over year. The most important contribution of the EU roadmap has been to zoom into definitions of "green" and "environmentally sustainable" economic activities in the form of a classification system. This system helps assess the strategic alignment of existing product and service offerings and procurement relationships with suppliers vis-à-vis peers operating in similar sectors and, ultimately, may help reach a common ground for a net zero carbon scenario.

DIRECT VERSUS INDIRECT ENVIRONMENTAL IMPACTS

A common roadblock that environmental sustainability practitioners find themselves struggling with is reconciling organizational commitments to a low (or lower) carbon-operating environment with the impacts that those targets may imply, given that corporate accountability is mostly defined by a multiyear timeline measured against a baseline year. Impacts may be direct impacts on the economic activities a business is involved in. From production, through the installed base of their operations to procurement decisions as relates to the selection of like-minded vendors, to people, both employees and consumers, and how the transition to low carbon affects existing health and safety standards. Or they may be indirect impacts – namely of the business as enabler of better environmental performance through policies fostering environmental literacy, or as a champion of sustainable practices in its sector and geographic markets, and through its B2B role with the distribution of environmentally conscious intermediate goods or services to other businesses. A roadmap of environmental sustainability that recognizes and addresses both direct and

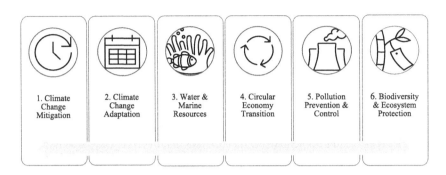

Strategic Focus #1. Climate Change Mitigation and Adaptation
Sample Sector: Agriculture and forestry
Sample Activities: Rehabilitation; reforestation; existing forest management; growing of perennial/non-perennial crops
Guiding Principles for Mitigation and Adaptation: Restoration and rehabilitation shall maintain and/or increase carbon sinks of above and below ground carbon:
- *Establish a verified baseline GHG balance of relevant carbon pools at the beginning of the rehabilitation/restoration activity; and*
- *Demonstrate continued compliance with Sustainable Forest Management requirements and increase of carbon from above and below ground carbon over time, supported by and disclosed through a forest management plan (or equivalent) at ten-year intervals, that shall be reviewed by an independent third-party certifier and/or competent authorities.*

Strategic Focus #2. Climate Change Transition/Enablement (Circularity, Pollution Prevention, Ecosystem Protection)
Sample Sector: Transportation and storage
Sample Activities: Freight transport services by road; inland passenger water transport; scheduled public transport
Guiding Principles for Transition/Enablement: Demonstrate substantial GHG emission reduction by:
- *Increasing the number of low- and zero emission fleets, and improving fleet efficiency; and*
- *Improving efficiency of the overall transport/mobility system.*

FIGURE 6.1 Environmental strategies: mitigation and adaptation vs. transition. Note: Activities listed are based on business performance or as enabler of improvements in other sectors. Author's intuition, adapted from the publication of the EU Taxonomy (2020).

indirect impacts of a company's environmental practices will position an organization to meet both shorter term and longer term climate commitments, and to translate them into lasting enterprise value that investors will be able to recognize and appreciate.

THE INVESTOR'S PERSPECTIVE ON THE CLIMATE COMMITMENTS OF BUSINESSES

As an investor, I have evaluated the impact of carbon emissions across a variety of domestic and international businesses. Some were publicly listed, while others operated regionally through a web of private subsidiaries. In gathering initial insights, I have discovered two main roadblocks: #1 data gaps so severe they would affect the validity of the analyses, and #2 the need to build a deep climate science expertise to be able to drive the analyses and interpret the outcomes.

The lens of financial materiality, specifically in line with the SASB guidance, allows investors to move past such roadblocks and deliver actionable metrics with both investment selection and risk mitigation in mind. Ultimately, the reason behind the evaluation of CO_2 emissions is to determine their impact on a company's economic and societal viability. That evaluation needs to map the spectrum of environmental events to the sources of "financially material" risk. When I first started my work on reconciling the financial health of a company with its resilience in light of environmental and social hardships, that was quite a novel approach. It was culturally not part of the perceived role of financiers and investors, whose legal obligations to clients are inherent in their delivery of positive financial returns for the risk taken. Saying this approach was largely dismissed by many is an understatement. It went from companies avoiding inquiries into sustainability risks to the outright marginalization of this approach as a potential threat of well-established and acceptable ways of doing business. No organization claimed to be the ultimate champion, and most pivoted that message as a short-term response to market pressures – whether customer demand or regulatory advances.

At the time of writing, sustainability risks are in the headlines every day, and sustainability-related opportunities are already creating the employment opportunities of the future. Financial capital is also being mobilized in large amounts toward impact-oriented goals. What has changed in only a handful of years? Perhaps climate literacy has not yet caught up with the many advances in sustainable innovation or with the science behind the most pressing environmental struggles. Yet, the underlying difficulty in progressing toward measurable impact goals and defining a believable aspiration for impact are far from being resolved in the financial sphere or among public policy officials.

I find that a tremendously valuable approach is to first define a series of intentional steps to frame environmental dimensions from the lens of direct and indirect impacts of the business that an investor is assessing. This is how I define it:

STEP 1 – IDENTIFYING A TARGET IMPACT

Use the financial materiality lens for a preliminary identification of carbon exposures:

- Define the sources of emission intensity at the asset level (either sector level or company level, depending on availability) to determine whether they involve operations (referred to as Scope 1), procurement activities (Scope 2) or distribution and supply chain (Scope 3).

- Prioritize the environmental impact we are looking to deliver: in this case prioritizing the direct environmental (E) risk of GHG emissions or the indirect E impact.

A *counterintuitive finding* that has emerged from my research is that when applying the financial materiality lens in prioritizing the environmental impact of greenhouse gas emissions, less than 20% of public companies in developed economies operate in sectors and value chains associated with direct GHG emissions (Scope 1) (Falsarone 2017). When we broaden the lens to include indirect E impacts that also carry CO_2 emission risk, such as the handling of waste and high toxicity materials or energy management, that number exceeds 50%.

Ultimately, focusing on direct, financially material exposures to GHG emissions by companies reveals only a small percentage of the entire ecosystem. Rather, I found a strong support for the argument that the impact of indirect activities on the E footprint of a company's operations had been intensifying given trends at the intersection of mobility and resource scarcity (which magnify that risk). Intuitively, it also means that if the focus on data is driven by the mapping of "green revenues" instead of GHG emissions risk, investors may risk capturing only financially *immaterial* data in terms of the long-term sustainability performance of the companies under review.

STEP 2 – MAPPING DIRECT EXPOSURES

Intuitively it seems that, over the past several years, investors' attention has been drawn to analyzing cumulative exposures by GHG emission intensity at the sector or country level. As carbon risk carries non-traditional financial consequences that are not easily priced before a negative or extreme loss is recorded, it requires prudent management. Moreover, while many large investors are advocating minimizing or avoiding exposure to emission-intensive sectors, that is not a guarantee of moving financial capital to a lower carbon future or delivering on their commitment to decarbonize their investments.

A *counterintuitive finding* is that, when defining GHG intensity at an individual company level, the non-energy sectors present the most GHG intensity contribution per company. In fact, while public energy companies represent approximately 50% of the market capitalization of all businesses with material exposure to Scope 1 emissions, the large number of companies in the energy ecosystem (over 80 at the time of the study) means that the average environmental impact per company is much lower than that contributed by more concentrated sectors, such as consumer cyclicals or basic industries (e.g., paper, chemicals or metals and mining businesses) because those sectors comprise fewer companies.

From the *investor's perspective*, this means that to minimize the E impact in terms of carbon emissions of the "usual suspects" – the GHG prone activities – the most impactful decision is managing sector-level exposures as opposed to exposures at individual company level.

STEP 3 – PROMOTING OPPORTUNITIES TO BUILD OUT A LOWER CARBON ECONOMY

The financial materiality lens is able to shift investor perception of environmental sustainability considerations from risk mitigation to opportunities by following a company's forward-looking trends. These are reflected in the quality of its management decisions and its environmental performance as compared with its competitors.

A *counterintuitive finding* at this stage is that the sectors that present a well-diversified number of companies to choose from (energy, for example) are also most likely to experience improving trends in climate risk when looking in aggregate at measures such as pricing of reserves adopted, capital expenditures devoted to the fuel economy, energy consumption and renewable footprint. This implies that investors can provide funding for the future energy transition of entire sectors of the economy in a responsible manner, instead of just excluding them from their pool of investments. (See Technical Note on Activity-Based Models and Input/Output Analyses as applicable to climate adaptation and resilience-building strategies.)

CIRCULARITY MODELS AND ENVIRONMENTAL STEWARDSHIP

Whether in a lecture, in the workplace or at a technical conference, I have found that when I talk about culture and the environmental footprint of businesses, listeners immediately think of "remediation liabilities" that organizations must face (or avoid) through the prudent management of their operations and products. Not surprisingly, environmental sustainability practices have historically been the realm of health and safety guidelines, which must fall in line with standards advocated by industry associations and comply with mainstream regulatory guidance on technical remediation practices. An avoidance of risk mindset, as opposed to the opportunity of leveraging greener corporate practices to support innovation and create a market for sustainable business activities, has been the primary driver for business to adopt environmental consciousness.

Circularity models, which draw their applications from the concept of a "circular economy," are one way in which businesses are responding to climate disruptions and the impact on natural ecosystems. When applied to production or consumption activities, a circularity mindset entails a human and economic system where the lifetime of resources employed in production is extended, through processes and technologies, to enable a recover-reuse model. Advances in digitalization have enabled vast applications of circular design models. Those models range from sustainable sourcing and management of resources through traceability, to optimizing waste collection and recycling programs that involve, among other things, the renewal of critical raw materials (e.g., those in scarce supply that lack immediate substitutes and hold significant economic importance) and the drastic reduction of plastic use. Table 6.1 provides contrasting examples of linear and circularity models in the natural rubber market.

TABLE 6.1

Linear vs. Circular Economy

Test market: As a commodity, the example of natural rubber serves well to review adaptation activities (i.e., cost of doing business) as well as transition enabling activities (i.e., technological innovation)

Linear economy paradigm:	**Adaptation activities** that balance off the linear economy paradigm in the natural rubber market:
	Take: *Sourced from plantations in Southeast Asia*
	Make: *Tires*
	Use: *30% natural rubber: 70% synthetic rubber*
	Dispose: *High dispose rate (end of life)*
Take > Make > Use > Dispose	**Pollute:** *During decomposition, rubber leach releases chemicals in soil and water sources*
	Social Cost of Adaptation: Dependent upon local government subsidy of small landholders
Circular economy paradigm:	**Transition enabling activities** that amplify the circular economy paradigm in the natural rubber market:
	Make: *Tires*
	Use: *30% natural rubber: 70% synthetic rubber*
	Reuse potential: *Flooring applications*
	Remake potential: *Limited (biotic waste subject to contamination and safety risks)*
Recycle > Make > Use >	**Recycle potential:** *low recycling input rate (end of life tires)*
Reuse > Remake > Recycle	Social Benefit of Transition: Dependent upon economic development of landholders; responsible plantation; fluctuations in commodity pricing

Source: Author's assessment. For additional insights: Critical Raw Materials (CRM) Alliance (https://www.crmalliance.eu/natural-rubber).

The cultural shift behind the design of circularity models is central to modern environmental stewardship. It continues to be instrumental in moving a business away from a remediation-only mindset to becoming a testing ground for innovation and commercialization of industrial processes that directly fight climate change.

One area that continues to have wide applications for circularity is that of water management. Novel designs for water and wastewater management processes rely on reverse osmosis systems, which make it possible to recover nutrients (such as phosphorus) and reuse wastewater sludge as feedstock.

Ironically, solving the undesirable effects of climate change may not necessarily drive a full acceptance of the existence of climate change or its causes. We may be able to solve the problem of climate change by changing its perception and focusing on adaptation strategies, instead of focusing on transition as a binary choice: Adaptation or Transition? Circularity models represent the way in between: Companies can move away from an environment in which risk avoidance and remediation liabilities force them to adapt to the inevitable outcomes and costs of doing business and move toward becoming enablers of low carbon transition technologies. (For an in-depth review of the value of circularity models for sustainable transitions of businesses, please reference the Technical Note in Chapter 10.)

Environmental sustainability practitioners find themselves reconciling organizational commitments to a low carbon future with measures of corporate accountability that are mostly defined by multiyear progress against a baseline year. As business models redefine environmental scenarios through the lenses of either mitigation or adaptation plans, building inclusive networks to shape those plans is vital.

What is the role of organizational inclusion in resilience planning?

Are there any analytical tools that can help a business measure the value of inclusion when setting sustainability outcomes?

THE LEARNING JOURNEY – CULTURE AND CLIMATE CHANGE

The transition of business models to either environmental adaptation or mitigation challenges organizations to recognize both direct and indirect impacts of climate change. The following checklist helps the reader identify climate-aligned business activities and to experiment with circularity models for an organization of choice:

- Identify the company's organizational priorities with respect to climate change (stated, long-term climate commitments versus intermediate-term operational targets). Apply a business and a multi-stakeholder lens to Adaptation, Mitigation, Transition and Enablement of climate-aligned activities to best describe the value attributed to each within the organization.
- Define the level of cultural awareness for environmental priorities within the business. List corporate activities that best describe a state of "urgent action" toward a climate adaptation strategy versus those that reflect a lack of cognitive acceptance and mobilization of efforts within the organization.
- Apply the Adaptation versus Transition lens to the direct and indirect environmental impacts of the sector the business operates in. Define the catalysts for change in the shorter and longer term. What value would an adaptation scenario bring to the sector? What if climate transition activities were instead central in a forecast scenario? What would have to happen today for your organization to lead in either scenario?
- Describe the role of circularity within the operational ecosystem of your organization. Can the Adaptation and Transition scenarios you investigated earlier be rendered as a circularity strategy? List the enablers

(e.g., technology advancements, policy and regulatory changes, investors' pressure, competition in the marketplace and/or change in customer preference for your company's products or services).

- Sketch an activity model to describe the systemic enablers that are at play for your business in terms of its environmental sustainability. Where could microlevel mapping contribute to the internal dialogue?

TECHNICAL NOTE – ACTIVITY-BASED MODELS OF CARBON REDUCTION POLICIES

Activity-based models describe human activities and decision-making over a timescale under the assumption of continuous repetition. Let us explore, for example, climate adaptation policies for the purpose of urban transportation planning with the goal of balancing both efficiency (transportation demand) and environmental outcome (carbon emissions created by human decision-making on type of transport, distance traveled and means of transportation available). By employing simulation analysis to transportation scenarios, we can assign a measure of *environmental value/benefit* to one transportation policy versus another (e.g., car-free streets vs. flexible commute times). This is done under the hypothesis of rational decision-making and is based on historical studies conducted in a variety of urban areas and geographies.

Carbon emissions, mostly generated from fossil fuel combustion, are the biggest contributors to greenhouse gases in urban areas and directly linked to economic activities carried out by urban populations. Setting carbon reduction targets in cities, for example, requires integrating transport activities surrounding urban areas as well as the economic activities carried out by businesses. Historically, the relationship between environmental health and transport has not been studied in the context of carbon reduction policies due to the complex data and comparability issues. For example, researchers have collected observations of air pollution levels in and around main transportation hubs. Such studies have focused primarily on establishing a baseline for human health and air pollutants, instead of planning of human activities (such as transportation/traffic patterns) and the response (changes in traffic behavior of users) to carbon reduction policies affecting traffic patterns and urban planning by the users (commuters).

In a direct activity-based model that simulates *transportation as a human activity*, individual behaviors and decisions toward transport and traffic need to be analyzed beyond distance metrics (e.g., the linear connection between start and end of travel). Model inputs may need to take into consideration the entire spectrum of decisions during transport time, such as: stopping points and time spent at each intermediate destination, and/or number of road segments traveled. It becomes clear that models focused on the microlevel would detect transport behaviors and would need to rely on data-intensive simulations of *need for transport* and *enablers of activity* in the model.

The chart in Figure 6.2 reveals how system thinking is vital in designing and implementing any activity-based modeling for carbon reduction policies. This is regardless of whether the initial scenario takes individual activities as core, or

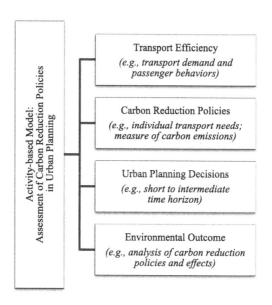

FIGURE 6.2   Microlevel mapping – transport activities. Author's intuition.

shifts focus to an entire organization or the economic systems and geographic context in which the business operates.

WHAT CRITICS MAY SAY

Over the past two decades, predictive models such as these have employed micro-simulations to describe traffic demand in cities such as Portland or San Francisco in the United States, as well as Suwon in Korea. What critics may say is that individual studies have different objectives. The selection of the simulation model reflects those objectives, as it involves representing travel route and distance and speed of each vehicle to estimate the carbon emission profile of each.

If we step back and shift attention to the built environment and the economic activities that drive transport needs in the chosen geographic region, we may broaden that perspective. Instead of looking at the individual commuter and the variables that drive his/her *utility function* when choosing one mode of transport over another, we can look to a company-level preference for opti-mizing its environmental footprint by adopting a strategic decision on how to implement carbon reduction policies for commuters. Here we continue to find potentially diverging objectives that company-level environmental data present. This supports the reason why most company-wide incentives to lower trans-port traffic have moved the commuter choice/means of transportation decision to the individual (the employee, the local transport authority for public transit schedules/frequency of service or for pick-up/drop-off of deliveries during busi-ness hours).

A word of caution on company-level environmental data used to evaluate internal policies versus the climate commitments of a business may, in fact, lead scientists, policy advisors and investors to step back and widen the magnifying lens of activity-based models to broader economic sectors (their demand and supply value chains). Company-level reporting of environmental impacts and the carbon intensity of operations is challenging. For this reason, activity-based models that rely on sector-specific economic activities should offer a better reference to test companies' environmental stewardship plans and the carbon reduction policies they employ. Nevertheless, as in real life, economic and financial data and carbon analyses are not synchronized. Notably, companies can restate prior years' sustainability information, causing an annual mismatch, while public companies report financial information on a quarterly basis. The more granular company-level assessments are, the more the quality of findings may be affected by the data collection efforts.

As reporting of company-level carbon emissions is not mandatory in most countries, the success of activity-based models to drive carbon policy decisions will depend on a couple of factors. One is the continued evolution of verification and assurance standards around carbon accounting in jurisdictions that mandate yearly reporting. Another is how to set the boundaries of analyses and assign accountability for emissions at the supply chain level, as organizations that operate across sectors rely on third-party providers of intermediate goods and services or of natural resources in their procurement process (also known as Scope 3 emissions).

REFERENCES

Bajic, A., R. Kiesel, and M. Hellmich. 2021. "Handle with Care: Challenges and Opportunities of using Company-Level Emissions Data for Assessing Financial Risks from Climate Change." *HEMF Working Paper*, 21 February 2021. doi:10.2139/ssrn.3789928.

Chung, U., L. Mack, J. I. Yun, and S. H. Kim. 2011. "Predicting the Timing of Cherry Blossoms in Washington, DC and Mid-Atlantic States in Response to Climate Change." *PLoS One* 6 (11): e27439. doi:10.1371/journal.pone.0027439.

Dietz, S., B. Bienkowska, D. Gardiner, N. Hastreiter, V. Jahn, V. Komar, A. Scheer, and R. Sullivan. 2021. "TPI 2021 State of the Transition Report." *Transition Pathway Initiative.* https://www.transitionpathwayinitiative.org/publications/82.pdf?type=Publication.

European Commission. 2020. "EU Taxonomy for Sustainable Activities." (2020 update). https://ec.europa.eu/info/law/sustainable-finance-taxonomy-regulation-eu-2020-852_en.

Falsarone, A. 2017. "ESG Integration Insights – Omnibus Edition, Managing the Carbon Footprint of High-Quality Bond Portfolios." Sustainability Accounting Standards Board.

Kim, J., S. Lee, K. Kim, and D. Jung. 2020. "A Study on the Process of a Carbon Reduction Policy Using an Activity-Based Model." *Chemical Engineering Transactions* 78: 193–198. doi:10.3303/CET2078033.

Marshall, G. 2014. *Don't Even Think About It: Why Our Brains Are Wired to Ignore Climate Change.* New York: Bloomsbury USA.

Sakurai, R., S. K. Jacobson, H. Kobori, R. Primack, K. Oka, N. Komatsu, and R. Machida. 2011. "Culture and Climate Change: Japanese Cherry Blossom Festivals and Stakeholders' Knowledge and Attitudes about Global Climate Change." *Biological Conservation* 144 (1): 654–658. doi:10.1016/j.biocon.2010.09.028.

Sasaki, N., and F. E. Putz. 2009. "Critical Need for New Definitions of "Forest" and "Forest Degradation" in Global Climate Change Agreements." *Conservation Letters* 2: 226–232.

SASB. 2020. IIRC and SASB Announce Intent to Merge in Major Step towards Simplifying the Corporate Reporting System. Media release, 25 November 2020.

von Carlowitz, H. C. 1713. "Sylvicultura Oeconomica, oder Haußwirthliche Nachricht und Naturmäßige Anweisung zur Wilden Baum Zucht." Leipzig.

7 Building Inclusive Networks

Across nearly all social sciences, and the field of business management, the word *inclusion* is often used as a synonym for *participation*. As a result, when evaluating the effectiveness of inclusion metrics, practitioners frequently employ exclusion criteria – either qualitative or quantitative – to identify a target population. In applied research, a similar line of thinking has been extended to *inclusive networks*. An example is building designs for inclusive schools and cities, education centers or transport hubs, which may consider safety and accessibility from the perspective of people with disabilities.

In recent years, the launch of financial literacy apps has helped automation play the dual role of both educating and providing low-cost access as a way of building inclusive financial systems. We may expect a similar trend to take place in the post-COVID-19 recovery. Disaster relief action plans will need to be built with well-defined inclusivity parameters and systems of intervention in place to prevent at-risk populations from being left behind. Access to clean sources of electricity is likely to offer additional examples of the need to design systems according to core inclusivity criteria.

Why is an inclusive mindset important when designing organizational initiatives and commitments around sustainability-oriented projects? Because the most effective outcomes are the ones that are scaled through innovation and the extension of technological advances to areas outside of traditional use cases of software design or a lab-scale hardware environment.

What we have here is a unique perspective – building use cases for technology solutions that, rather than addressing how to optimize cost per user or other metrics of repeated visit in a user-oriented domain, look to identify an outcome that extends beyond the original target population and therefore encompasses all the beneficiaries of societal advances.

The growth of social media platforms, digital marketing and open architecture design allows one piece of information (a data field) to reach many more users than a targeted advertising campaign. Word of mouth, delivered digitally through a user's multiple networks, has created a situation where target population size is not as meaningful in a social impact study as the feedback on the action taken by first, second or third-degree users of that information.

The proliferation of ecosystems of developers reflects that catalytic, nearly invisible effect of innovating in an open space, where voices and inclusion are not planned nor measured. It just happens. Historically, organizational tools that tackle inclusive networks and catalyze more transparent tools of inclusion have relied on user experience as a metric of success. User experience may still be relevant when tackling

DOI: 10.1201/9781003212225-8

organizational inclusion and building internal efforts for a business to deliver credible and verifiable sustainability targets for societal impact. But the use of *experience metrics* to address the level of employee or other stakeholder support for such targets is best framed in the context of building an internal trust system.

To be successful, inclusive networks that catalyze internal efforts for impact must amplify stakeholders' participation and create a movement with values that are aligned to corporate sustainability commitments.

The topic of energy transition offers a concrete example of the widespread need for businesses to think about how to best organize inclusive networks to spearhead the transition of their economic activities through participation, whether those activities are in manufacturing, sourcing or distribution. Both scientists and financiers have cast their preference on how to deal with the technical needs of shifting industrial processes from fossil fuels to renewable energy. But the social backdrop of the transition is full of misaligned perspectives and diverging solutions on the "how to" of organizing our low carbon future. There are three tenets to bear in mind:

1. environmental justice considerations,
2. adoption of those technical solutions by communities that operate in a wide range of economic development, and
3. social policy actions to aid the practical shift as much as the behavioral one.

Scenario analyses are powerful tools that can address social factors in their multidimensional profile and ensure inclusivity when defining alternative energy pathways. They are useful in several ways. The ability to capture both environmental and socio-economic impacts as they evolve on a timescale is core to scenario-based analyses that deal with energy transition. Scenario analyses are also particularly useful in spurring an active, multistakeholder dialogue on societal shifts in the deployment and consumption of energy products. Finally, scenarios can incorporate feedback mechanisms through participatory learning and the short-term versus long-term impact of technology adoption on socio-economic variables. (See Technical Note for a discussion of the Delphi Method.)

Inclusive networks foster the creation of trust environments. In the context of the impact challenge for businesses, an example is group settings where company commitments are met by open and transparent decision-making. Analytical tools that promote inclusivity, from input gathering and data collection to the analysis, use and sharing of information, are central to lay out all possible impact scenarios for businesses and civil society in an environment of trust and transparency.

The role of scenario analysis and participatory learning is even more important in the context of the much-awaited release of a social taxonomy in Europe. With the release of the EU's taxonomy regulation in 2020 (the EU Taxonomy), businesses and investors have access to minimum thresholds laid out as guidance to distinguish between activities that advance the principles of sustainability and ones that do not.

Notwithstanding criticism, the EU Taxonomy has quickly become the reference framework for international policymakers, specifically when addressing which economic activities to deem not harmful to our planet. A year later, in 2021, the European Commission has begun to extend its work. It now incorporates multistakeholder dialogue on social criteria that may pose trade-offs in achieving environmental benefits and uses a human rights lens to identify activities that have the highest social value (a social taxonomy).

The roadmap adopted in the development of the EU social taxonomy has repercussions. It can potentially make it possible to legislate minimum criteria for acceptable environmental and social sustainability standards for business and investment activities. For this reason, it is likely to introduce organizing principles to ensure inclusivity is integrated into the social and environmental fitness of businesses. When balance is sought between human rights, governance of business activities and promotion of adequate living and working conditions, the identification of *social minimums* relies on international norms rather than science. Ironically, this is contrary to the foundation of the green eligibility criteria in the EU Taxonomy regulation. The development of social minimum criteria to support the environmental and societal commitments of businesses and their investors will continue to be a key area to watch when identifying the socio-economic variables that underpin the formation of inclusive networks in the workplace and beyond.

HOW TO ELICIT INCLUSIVITY IN A NETWORK

When it comes to gauging the level of inclusivity that business leaders are using in their decision-making, it is helpful to reframe it from the perspective of the participants. By taking *inclusion* as a synonym for *participation*, it becomes clear how this can be both a process and an outcome. We can borrow from the EU social taxonomy, which makes reference to the AAAQ framework (Availability, Accessibility, Acceptability and Quality), to define the flavor of strategic participation that will drive optimal outcomes for a business.

The AAAQ framework helps address the two key pillars for successfully gauging how inclusive a business decision is.

Pillar #1: *How can organizations create a set of tools and analytics that align internal efforts with corporate ambition and reward behaviors that catalyze resources to deliver those commitments?*

Pillar #2: *How can businesses benchmark their own sustainability goals and aspirations while driving organizational culture to a place of inclusivity and innovation?*

International organizations such as the United Nations and the World Bank have spent decades working to establish measures and analytical approaches to address social inclusion as a means to make their efforts lasting and more impactful. It turns out that those tools are applicable to private businesses as well, regardless of their size or sector.

Separating participants from nonparticipants by their identity features (e.g., ethnicity, gender, disability status, location, employment status) defines the "excluded groups" as well as the area of exclusion (by AAAQ). However, these measures are often useful to identify what may be termed *symptoms*, while the drivers of exclusion and the process behind which exclusion is produced are far more important.

One of the best laid out applications of the value embedded in inclusive social networks in the area of Accessibility and Availability was carried out by scholars from the Institute of Computing, University of Campinas, Brazil, in their e-Cidadania research project funded by the Fundação de Amparo a Pesquisa do Estado de São Paulo (FAPESP) and the Microsoft Research Institute (Table 7.1). The authors used a workshop setting to elicit stakeholders representing Brazilian civil society and clearly articulate the point of view of each, from those digitally excluded and in the lower end of the education spectrum to the ones building competencies and skills. This defined the context in which decision-making regarding Accessibility and Availability occurred. It was then possible to collaboratively

TABLE 7.1

Sample Evaluation Framework and Findings : The e-Cidadania Project

	Questions & Problems (when establishing an inclusive network)	Ideas & Solutions
Contribution	Participants **Technical Access | Ease of Interaction**: Education Level (literacy, digital proficiency) Promotion of universal access **Mindset**: Resistance to new; privacy issues; ethical use	Audiovisual Popular language User-friendly System alerts Ease of misuse reporting
Source	Clients | Suppliers **Sustainability of Network**: Longevity of participants in the system Dissemination of progress Identification of adjacent communities	Popular use/service Collaborative web-based interactions Elicit input for service/system evolution
Market	Partners | Competitors **Quality and Cost of Maintenance**: Complementary tool to existing offerings/services Focus on average user accessibility/features	Collaborative initiative Sponsors Open, transparent input Social interactions
Community	Observers | Policymakers **Competing Commitments**: Existing initiatives/offerings by NGOs Community awareness vs. balanced information flow Define a dynamic timeline and outcomes	Promote interaction with non-participants Articulate common interests and divergent opinions Highlight core vs. satellite problems

Source: Author's own interpretation of the e-Cidadania Project. For illustrative purposes only.

envision a solution, discuss both opportunities and consequences for each sce-nario, and directly address the socio-economic and technological divide between the interest groups.

The researchers relied on organizational artifacts such as Stakeholder Analysis Charts and Evaluation Frameworks to create a participatory approach to the prob-lem, along with key principles of the discipline of Organizational Semiotics.[1] By designing the problem in this way, the e-Cidadania study shows how to establish a mode of communication, stemming from the input received from a wide vari-ety of participants and subsequently processed through computer interaction that translates into a set of nontrivial, actionable priorities.

THE RISE OF SOCIAL ANALYTICS

Over the past few years, there has been an unexpected rise in awareness surround-ing the vital role that basic social dimensions – such as health and safety hazards, employee wellbeing and accountability toward key human rights, to name just a few – play as direct operational risks to the longevity of any enterprise. In particu-lar, technology solutions trained by adaptive AI to detect bias in the workplace are likely to rise to the top of the wish list of both Chief Risk Officers and Chief Technology Officers over the next handful of years. How can we best prepare?

THE PATH TO IDENTIFYING SOCIAL BIASES STARTS FROM WITHIN

The year 2020 may as well be referred to as the year of social awakening. As COVID-19 continued to infiltrate every aspect of our daily lives, it unveiled the fundamental weaknesses of our social fabric. In a matter of weeks, the #BlackLivesMatter movement pressured thousands of organizations internation-ally to revisit their operating practices. How could they better prevent and mitigate systemic inequalities to contain social unrest and build support for antiracism advocacy? Equal representation by race and gender, strength of community rela-tions and corporate behaviors toward basic human rights abuses were just a few of the social dimensions in question.

What became abundantly clear was the need to move the focus from stan-dards of conduct to defining an enterprise-wide toolkit for educating and guid-ing all stakeholders – both internal and external. While defining prevention and monitoring strategies sounds like the most sensible near-term approach to managing safely through social turmoil, it may quickly turn into the least effec-tive solution.

The hidden price of having a socially disconnected talent pool is faltering employee productivity, followed by the inevitable hindering of innovation caused by social biases in team dynamics. This is something that businesses were not immune to before 2020. The May 2019 Facebook report "Deskless not Voiceless: Communication Works" surveyed remote workers from over 4,000 companies with over 100 employees in the US and the UK. Fifty-four percent said they felt disconnected and "voiceless." Only 20% thought their ideas made up a substantial portion of conversation with their managers. Interestingly, the report predicted that three in four small businesses would have remote working arrangements in place by 2028. A future that the pandemic has significantly amplified.

DEVELOPING SOCIAL ANALYTICS
TO ADVANCE WORKPLACE EFFECTIVENESS

The "measure what you manage" approach to human capital has its critics. Long-time opponents often blame the lack of comparable datasets and privacy barriers for the minimal disclosure standards adopted by corporations, which makes social metrics too weak to drive corporate decision-making. This is a sensible explanation, which nevertheless may expose organizations to longer term issues and reputational struggles. The full-time home confinement and teleworking days of COVID-19 quickly made the impact of physical and mental health on operational performance a top priority in business continuity planning.

A recent study by McKinsey & Co. found that, in addition to basic needs such as safety and security, what is having a disproportionate effect on our newly digital workforce is the interplay between social cohesion, individual purpose and trusting relationships. The value of collaboration tools in strengthening individual and group interaction has been long recognized by business leaders, but there continues to be a huge gap in technology adoption to validate that. In fact, while almost 95% of leaders have identified the need for collaboration tools, only 56% currently use them. In addition, most of the adoption effort pre-COVID addressed workplace connectivity, not employee experience.

SOCIAL INNOVATION CAN PROPEL INCLUSIVENESS
IN HUMAN INTERACTIONS

Social innovation can help close the inclusion gap by making group communication collaborative and unbiased. An example of real data supporting change in real dynamics is that of the AI-enabled team communication platform RiffAnalytics. ai. Backed by an all-star crew of MIT-trained funders and advisors, the platform provides feedback on team meeting dynamics and promotes actionable insights. It does this by producing metrics of interruption/flow, dominance, bias and influence in a discussion, while preserving the privacy of human interactions.

Beth Porter, CEO of RiffAnalytics.ai, believes there is value in data that backs up human observations. Using tools that objectively measure how teams work together helps them to work more effectively and relate better to one other. Moreover, this data does not have to be delivered through managers. It can be sent directly to individuals and teams, allowing people to self-manage.

There is tremendous room to grow for human-centric solutions that enhance the well-researched productivity gains that stem from a connected workplace invested in employee engagement and inclusivity (on average a 17% increase in profitability as reported by Gallup in 2019). While it is impossible to predict the many ways in which the outlook for machine-to-machine communications will evolve over the next decade, the pick-up in mass connectivity will have to go beyond digital readiness and address the interplay of social biases and human capital development by placing employee engagement at the heart of technology adoption.

INCLUSIVITY AND BUSINESS OWNERSHIP: THE CASE OF OPTIMAX SYSTEMS

The intersection of employee wellbeing, organizational incentives and business ownership is an area I have continued to delve into as a fellow in the Business and Society Program at the Aspen Institute, a global nonprofit organization committed to realizing a free, just and equitable society. Building an environment where trust is a foundational lever of corporate accountability and societal purpose is deeply connected with the business ownership structure that an organization chooses.

Rick Plympton, an Aspen Institute Fellow in the Job Quality Program, shared with me the example of his pioneering work as CEO of precision optical components manufacturer Optimax Systems, Inc. Founded in the early 1990s, Optimax is leveraging research from the University of Rochester that was funded by Kodak and Texas Instruments to manufacture precision optics with computer-controlled machinery. Under his leadership, in 2020, Optimax adopted an innovative corporate structure: That of an Employee Ownership Trust (EOT). The EOT marked the transition of the company from private ownership to a perpetual purpose trust as ultimate owner of the commercial enterprise. It was the first of its kind in the US manufacturing sector. In legal terms, that meant creating a for-profit model, owned by a trust and led by a purpose statement comprising three key tenets:

1. Don't sell the company;
2. Continue to share monthly profits with the employees;
3. Ensure leadership is fostering innovation and the creation of new jobs to meet emerging market needs and for the benefit of the Optimax workforce and their community.

An independent board of trustees is responsible for ensuring senior leadership is driving business to enhance industry innovation and economic growth and new job creation in the regions where it operates.

Optimax is an example of how purposeful ownership and organizational incentives that are aligned with a holistic mission can redefine the culture of an organization. The reason why a corporate structure that enables employee ownership today, not tomorrow, was chosen is the foundation of the alignment of incentives that the leadership team at Optimax envisioned. One of personal accountability and organizational purposes. The highlights of my discussion with Rick speak highly of the sheer authenticity that guided the adoption of inclusive metrics of business success at the very core of business ownership at Optimax.

I asked Rick how he measures the power of building inclusive networks in his employee base. Did he target higher retention or other measures of wellbeing? Rick answered that his vision was to build a lasting enterprise that thrives locally and shares wealth at the local level while continuing to grow organically. Corporate culture established over the years has helped foster an environment where transparency is respected and expected from employees.

Rick and his team hold a monthly meeting to update the workforce on the company's profitability. Each month, profits are allocated according to a 25:25:50 rule: 25% is paid to employees, approximately 25% is set aside to pay taxes, and the remaining 50% is reinvested to grow the business. By bringing full transparency on the financial standing of the company, all employees – regardless of their tenure – take part in the profit sharing. The bonus plan is structured such that employees are fully vested after five years of employment and earning US$50,000 per year; to put a finer point on it, the janitor, the R&D engineer and the president all get the same monthly bonus check – which has averaged about US$1,000 per month for the past several years. In addition, in recent years the company has established a retirement plan with matching contribution by the employer and 20% of the bonus income is allocated toward retirement. The profit sharing and 401k plan provide the opportunity for every Optimax employee to retire as a millionaire.

Optimax's ownership structure is highly unusual, and I also wondered how Rick came to realize that employees would be in favor of the change in corporate ownership structure. I discovered that the discussions started at a mature stage when the business had already proved to be on sustainable financial footing. By virtue of having adopted a fully transparent communication model between business owners and workforce, the decision to grant employees access to profit sharing was embraced without diffidence or confusion. Optimax leadership had built an environment where trust is a foundational trait of its organizational DNA.

I also found out that the ultimate outcome for the company was local economic development. Rick has been active in the local business community as well as nationally in sharing the Optimax journey as the first US-based manufacturing company to embrace an EOT model. The challenge he will confront next is how to ensure retirement savings and long-term employee wealth is invested responsibly and in a manner consistent with the goal of growing wealth and contributing to economic development locally. To support regional infrastructure plans and build resilience at the municipality level, Optimax has actively partnered with financial institutions to create innovative investment vehicles as alternatives to traditional retirement plans.

Organizational inclusion helps address both financial and societal challenges when setting sustainability outcomes. Social analytics and new business ownership structures promote inclusivity and can play a central role in building an environment of trust and transparency.

How can organizations promote the adoption of social innovation within their ecosystem?

Do group activities inspired by elements of gamification help getting past potential roadblocks to adoption?

THE LEARNING JOURNEY – BUILDING INCLUSIVE NETWORKS

By leveraging participatory techniques, the following checklist encourages the reader to apply the science of inclusion to promote company-wide participation in delivering impact goals:

- Through the lens of participatory learning techniques, determine which metrics of inclusion are relevant within your organization and whether potential exclusion criteria may have been in place.
- Take note of those tools that have historically relied on "user experience" as a metric of success.
- Leverage experience metrics to address the degree of multistakeholder support of sustainability goals to build an internal system of shared trust and transparency. Highlight the knowledge gap that external experts could fill.
- Drawing insights from the set-up of the Delphi Method, sketch the areas that external expert questionnaires should cover in the first and second round of engagement. Take note of the internal forums that would be best suited for the findings to be shared with.
- Refer to the AAAQ framework (Availability, Accessibility, Acceptability and Quality) to define the optimal level of strategic participation that will drive an impactful outcome for the business.
- Single out which organizational initiatives are best aligned with the sustainability commitments that are sought by the business while optimizing for the pillars of the AAAQ framework.

TECHNICAL NOTE – THE DELPHI METHOD

The Delphi Method is an analytical process for forecasting and prediction analytics, which relies on the structured integration of dynamic feedback from a panel of experts. Also known as the Estimate-Talk-Estimate (ETE) technique, it has been employed in a wide range of applications for business forecasting (e.g., industrial automation, clinical trials) as well as policy settings (e.g., health and education, e-democracy).

Expert opinions are incorporated in the forecasting exercise through repeated questionnaires, where forecasts and reasons for the judgment expressed in each answer are shared in an anonymized form among panel participants during multiple rounds. The principal assumption of the Delphi Method is that group input on an issue is more powerful than any of the individual judgments. To produce a stable outcome (one which exhibits limited variability), the ETE process is mediated by a facilitator who helps the group converge to a consensus answer to the problem under investigation. Each panelist can alter their input in real time and can provide feedback on other participants' views anonymously. To avoid methodological biases, panelists are not informed of the group composition either during or after the process.

AN APPLICATION OF THE DELPHI METHOD: INCLUSIVE NETWORKS

The selection and analyses of environmental commitments by a business are presented in the context of their implementation within the operational processes already in place. It leverages the broader framework by Gallotta and Garza-Reyes (2018) to implement any type of sustainability initiatives in multifunctional and intradepartmental settings as referenced in the Proceedings of the International Conference on Industrial Engineering and Operations Management. The complexities in environmental impacts that businesses are exposed to in carrying out their standard operations are reviewed. This is done both in terms of their physical impacts associated with climate risk, and their need to adapt to changing climate scenarios in their localized operations and their dispersed supply chains (Figure 7.1).

In this example, the question we aim to answer by employing the Delphi Method is:

Which environmental targets reflect the highest priorities for a business that is looking to amplify both its societal and financial value?

When designing the first survey to solicit input from the expert network, I define three principles to guide the first round of assessments:

1. **Accountability**: Perceptions, industry best practices and organizational alignment to best identify the spectrum of environmental factors (EF) that affect the company.

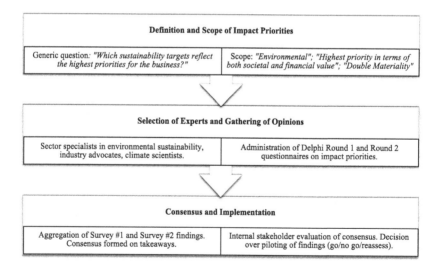

FIGURE 7.1 Problem set-up in the Delphi Method. Survey analysis adopts a Likert scale (1–5, from strongly agree to strongly disagree) as traditionally utilized in psychometric assessments administered digitally. Consensus is achieved when a threshold of 75% (three quarters) reaches agreement in the response analysis of Round 2 (Survey #2). Author's own intuition.

2. **Influence**: Internal and external lens on relevance of EF on the existing business scenario versus future developments at the company or industry level.

3. **Maturity**: Time needed to implement existing sustainability initiatives addressing EF and urgency in terms of the need for future time commitments to support implementation of environmental targets and reporting of outcomes.

As Round #1 closes and is analyzed and discussed through anonymized feedback by the experts, Round #2 introduces principles of **Readiness** (discussion of existing metrics vs. future progress), **Dynamism** (organizational commitment to push progress forward by setting new targets) and **Transparency** (willingness to openly communicate the sustainability journey of the business in a programmatic way). This sets the stage for refining the first layer of outcomes. Round #2 introduces a component of pragmatism and the existing need for additional resources (time, people, capital) to support the delivery and communication of environmental outcomes in line with a range of operating scenarios and existing and future targets (Table 7.2).

TABLE 7.2
Sample Survey Design

Round #1	Round #2
Principle: Accountability	*Principle: Readiness*
Should EF be deemed material for a diversified industrial manufacturer?	Should metrics adopted to monitor material EF be quantitative or qualitative?
Should EF be discussed in the context of business risks/reputation, opportunities, both?	
Should EF be part of the growth strategy of the company? (*)	Should they highlight gaps/progress made vs. a target?
Should full oversight of EF be the responsibility of one functional area? (*)	
Should EF be addressed separately by specific policies and compliance practices? (*)	
(e.g., health & safety; land contamination; green energy procurement, etc.)	
Principle: Influence	*Principle: Readiness*
Do EF affect existing business operations (internal factors)?	Should metrics describing material EF be dynamic or static?
Do EF affect business development and customer satisfaction (external factors)? (*)	
Do EF affect the wellbeing of employees? (*)	What is an acceptable time horizon for re-evaluation?
Do EF affect the ability to retain and recruit workforce? (*)	
Do EF affect procurement cycle and supply chain relationships? (*)	

(Continued)

TABLE 7.2 (*Continued*)
Sample Survey Design

Round #1	Round #2
Principle: Maturity	*Principle: Transparency*
Are EF being thoroughly addressed in product lifecycle (legacy and new products)	Should metrics describing material EF be publicly
Are EF associated with clear organizational incentives across the organization? (*)	disseminated or internally reported?
Are EF reviewed on a fixed timeline or addressed on an ad hoc basis? (*)	
Are EF outside of existing policies evaluated/monitored as emerging areas?	
Are EF communicated and discussed openly within the organization? (*)	

Source: Author.

Note: EF = Environmental Factors; Likert scale utilized in Round #1 Survey: 1, Strongly Disagree; 2, Somewhat Disagree; 3, Neither Agree nor Disagree; 4, Somewhat Agree; 5, Strongly Agree (*) Level of Inclusivity in driving EF outcomes.

WHAT CRITICS MAY SAY

A word of caution. The forecasting ability and accuracy of this method are highly dependent on three factors: (1) the panelists' level of expertise in the area under evaluation; (2) the ability of the facilitator to guide the process, as responses are aggregated and conflicting perspectives are evaluated, until consensus is reached by establishing appropriate thresholds; and (3) the time horizon of the forecast. The first applications of the method can be found in science and technology fields, where innovation advances follow development timelines that are hard to predict and even harder to express with a single indicator of consensus. Nevertheless, the introduction of web-based platforms to conduct surveys and questionnaires among expert groups is enabling a new stream of real-time ETE evaluations.

A second word of caution. The difference between the Delphi Method and the science behind prediction markets is that while they both rely on the diverse opinions of a group, in the latter case, participants self-select as opposed to being selected as experts by a facilitator. In addition, in a Delphi evaluation, the reason behind the opinion expressed is also made available under the condition of anonymity. In recent years, stock exchanges have developed a body of research where the setting of an ETE evaluation is combined with traditional prediction markets and a one data point input is substituted by a range of data points as forecast.

NOTES

1 Authors' definition of Organizational Semiotics (OS): OS studies the nature, characteristics, function and effect of information and communication within organizational contexts. Organization is considered a social system in which people behave in an organized manner by conforming to a certain system of norms. These norms are regularities of perception, behavior, belief and value that are exhibited as customs, habits, patterns of behavior and other cultural artifacts – See Technical Note in Chapter 3.

REFERENCES

Clarke, S., and N. Roome. 1999. "Sustainable Business: Learning–Action Networks as Organizational Assets." *Business Strategy and the Environment* 8 (5). doi:10.1002/ (SICI)1099-0836(199909/10)8:5<296::AID-BSE212>3.0.CO;2-N.

Conway, M., and R. Plympto. 2021. "Job Quality and Employee Ownership: An Interview with Job Quality Fellow Rick Plympton (CEO, Optimax)." *The Aspen Institute, Employment and Jobs Blog Post*, 11 March 2021. https://www.aspeninstitute.org/-blog-posts/job-quality-and-employee-ownership-an-interview-with-job-quality-fellow-rick-plympton-ceo-optimax/.

Duttal, P., and A. Kumaravel. 2016. "A Novel Approach to Trust Based Identification of Leaders in Social Networks." *Indian Journal of Science and Technology* 9 (10). doi:10.17485/ijst/2016/v9i10/85317.

Falsarone, A. 2016. "Delivering Mission-Driven Value: The PBC Balancing Act." *NACD Magazine* (May/June).

Falsarone, A. 2020. "Financial Innovation and the Rise of Social Analytics." *Center for Financial Professionals Magazine* 16 (July).

Gallotta, B., and J. A. Garza-Reyes. 2018. "Using the Delphi Method to Verify a Framework to Implement Sustainability Initiatives." *Proceedings of the International Conference on Industrial Engineering and Operations Management Bandung*, Indonesia, 6–8 March, 2018: 231–241. http://ieomsociety.org/ieom2018/papers/69.pdf.

Hayashi, E., V. Neris, L. Almeida, C. Rodriguez, C. Martins, and C. Baranauskas. 2008. "Inclusive Social Networks: Clarifying Concepts and Prospecting Solutions for e-Cidadania." *Technical Report* IC-08-09. Institute of Computing, University of Campinas – Brazil.

Liu, F., and H. J. Lee. 2010. "Use of Social Network Information to Enhance Collaborative Filtering Performance." *Expert Systems with Applications* 37 (7): 4772–4778.

Platform on Sustainable Finance. 2021. *Social Taxonomy Outreach*. 26 February 2021. https://ec.europa.eu/info/sites/info/files/business_economy_euro/banking_and_finance/documents/finance-events-210226-presentation-social-taxonomy_en.pdf.

Prokesch, T., H. A. von der Gracht, and H. Wohlenberg. 2015. "Integrating Prediction Market and Delphi Methodology into a Foresight Support System — Insights from an Online Game." *Technological Forecasting and Social Change* 97: 47–64. doi:10.1016/j.techfore.2014.02.021.

Revez, A., N. Dunphy, C. Harris, G. Mullally, B. Lennon, and C. Gaffney. 2020. "Beyond Forecasting: Using a Modified Delphi Method to Build Upon Participatory Action Research in Developing Principles for a Just and Inclusive Energy Transition." *International Journal of Quantitative Methods* (27 February). doi:10.1177/1609406920903218.

8 The Informal Organization

When a business chooses to embrace an environmental or societal hurdle that directly affects its operations or ecosystem – customers, peers, employees, geographic regions – what determines its chances of success? Regardless of what drives that company to embrace that challenge – whether it is sheer business interest, alignment with corporate purpose or regulatory ask – the way the organizational fabric responds determines the likelihood of its success.

In previous chapters, we learned to recognize culture as a key enabler of organizational learning around sustainability commitments for businesses. In fact, the *informal organization* – the relationships and mindsets that stir internal processes and people to influence the delivery of business value in a company's day to day – may move at a different pace from that planned as part of a well-crafted sustainability integration strategy.

Over the decades, I have retained one simple yet powerful takeaway from my graduate courses in operations research at MIT: "Processes are people." At first, that idea may suggest a lack of control over planned corporate program rollouts. But it brings attention to the need for dynamic observation of intermediate outcomes as the rollout progresses. "Processes are people" calls for a mindset that can live with more approximate answers and less definitive solutions. Yet it invokes a continuous learning environment and dynamic fine-tuning of accountability and outcomes in the organization's sustainability journey.

There are two core elements that make planning for sustainable outcomes most effective. First, when defining the sustainability commitments and targets of a business, placing *transparency* first in both internal and external stakeholder engagement. Second, leveraging the power of *gamification* to identify and build consensus for impact-oriented targets. This applies whether the target is related to environmental sustainability, which is usually dependent on a long-range time horizon spanning decades, or to social measures, which may entail shorter term action items.

Embracing "transparency first" as a mantra to inspire the informal organization and persuasive game design to elevate sustainability awareness sets a level playing field between corporate ambitions and the pursuit of sustainability commitments, while also maintaining an inclusive mindset.

THE REAL VALUE OF TRANSPARENCY

More often than not, transparency in a business context is associated with the set of norms and guidelines that an organization has chosen to guide the degree of public disclosure surrounding either strategy direction or the material business risks and opportunities it faces. Whether viewed through the internal or external stakeholder lens, a "transparent organization" is one that has been successful at keeping sources of *reputational risk* in check.

Explanatory Note: Practitioners and academics have rarely directed their analysis of the alignment of for-profit businesses to stated mission or core values as a transparency check. Nor have they tried to evaluate whether maintaining a degree of integrity to those values and mission may be an indicator of future business performance. A primary reason for this could be the lack of integrated information systems for both internal and external communication, coupled with the slow evolution of standardized disclosures of sustainability in business reporting. Even with the unprecedented international efforts and rapid adoption of comparable measures of social, ethical and environmental performance, verification and provision of reasonable assurance of those disclosures is still a work in progress.

As governance practitioners would attest, the *real* value of transparency is turning away from the negative connotation of reputational risk and shaping existing governance structures – those that historically have been placed as acceptable safeguards of a company's license to operate – into an accountability engine powered by inclusivity and cross-sectional innovation.

THE EVOLUTION OF REPUTATIONAL RISK

In June 2019, I was invited to give a keynote talk on reputational risk at the Annual GRC (Governance, Risk Management and Compliance) Summit in Baltimore hosted by MetricStream, a pioneer software provider in the space. The theme of the summit was "Performing with Integrity." I confess that in the middle of one of the busiest years I have ever had and the prelude to summer days, it was the inspiring theme that sold me on participating and stretching my energy further.

First, I addressed reputational risk in the context of an organization's traditional, enterprise-level risk management framework. Second, I challenged the group to ponder whether any new or emerging metrics should be considered, given the increasing push for transparency in corporate behaviors by financial and non-financial stakeholders. Lastly, I discussed how to assess whether the business may have done a good job managing its own reputational risk trajectory as opposed to relying on simple benchmarking to peers. Little did I know that the experience of giving that talk would influence the way I defined and valued the role of transparency in business and in life. Let us retrace that journey together...

The term "governance" is widely used in the day-to-day activities of corporations. But it is interesting to compare how the various definitions stack up against one another depending on the context. For example, it should be no surprise that, according to the CFA Society (the association that awards the Chartered Financial Analyst's certification), governance as a corporate function is defined as a system of controls that are put in place to benefit the institution, its management team and its so-called "governing" board of directors. With that in mind, it is also not surprising that governance systems may be influenced by interest groups with different or divergent interests.

On the other hand, the definition of reputational risk has expanded due to an increase in the desire to link corporate decision-making with organizational incentives and broad stakeholder commitments. Let us think, for example, of:

- Heightened regulatory demand for transparency vs. short-termism
- Scrutiny over pay4performance in executive compensation arrangements
- Deeper awareness of the role played by intangible variables such as diversity
- The rise of millennial consumers and millennial employees
- Technology reshaping virtually every business function

These factors are raising the integrity bar on every front. Reputational risk was once used to address *business readiness* – in other words, to answer the question: "Are we doing enough to build organizational resilience?" Now it is used to answer the questions: "Are we capturing everything when building resilience? Are we fully understanding the role played by non-financial variables?"

The change in inquiry that such questions entail shows us that, while remaining fully intertwined with governance decisions taken by leadership teams and boards, the evolution of reputational risk has drifted out of sync amid the rapid pace of change.

The rise of controversial business activities as monitored and reported by Swiss reputational risk and ESG due diligence provider RepRisk provided an important reminder to our group. Governance structures that fail to address the implications of emerging reputational risks are a strong indicator of *corporate commitments* (financial or societal) that are likely to be disconnected from day-to-day *corporate behaviors*. Yet those behaviors are the ones that the informal organization follows and values as "precedents" in decision-making. For example, an assessment of corporate complicity, with respect to repeated lack of transparency over lobbying disclosures by public companies, was the leading factor for a company being flagged as a violator of international anticorruptions norms and regulations.

Therefore, as new, more sophisticated facets of reputational risk continue to threaten business resilience, *transparency* remains a primary tool when aligning governance and accountability structures to societal challenges that are quickly moving from having intangible to tangible outcomes.

WHAT CROSS-SECTIONAL RISK EVALUATION
SAYS ABOUT TRANSPARENCY

During the five years to the end of 2019, I repeatedly surveyed 1,000 corporate practitioners and expert groups to capture how reputational risk was evolving in response to emerging areas of focus in sustainability. The study also aimed to capture how these emerging reputational risks might have been perceived across corporate functions by the *informal organization*.

Similarly to the set-up of the Delphi method, the five-question survey was run in a group setting, with on-site reporting of answers to the questions followed by group sharing and discussion. It helped me refine my own intuition about reputational risk being directly tied to potential failure in operational or governance processes – processes that are essential to diligently carry out corporate objectives. While the findings stress the relevance of translating reputational risk in the context of each functional area, they unambiguously point to reputational risk being a societal risk (based on today's insights about emerging trends), not a risk tied to the depth and transparency of company's disclosures of past operational and financial results nor depending on its status as a private versus public company.

The survey questions, and their findings, were as follows:

Question #1. In your experience, is reputational risk on an increasing, decreasing or steady trend within your organization(s)?
Finding #1. On an increasing path. Most participants attributed the trend to higher organizational awareness about the environmental and social responsibility of businesses and a lower tolerance of negative headlines by consumers and shareholders.

Question #2. Should reputational risk be categorized as "culture risk" or "operational risk" (e.g., a cost of doing business)?
Finding #2. Reputational risk is increasingly perceived as a culture risk yet addressed as an operational risk.

Question #3. Who is accountable for the management and oversight of reputational risk within the organization, for example the board, senior leadership, customer-facing functions, everyone?
Finding #3. The oversight and management of reputational risk belongs to everyone involved and does not just rely on traditional governance structures in which ownership lies with the CEO or the board of directors. An increasing number of responders thought the most widely accepted resolution was to "pay the monetary fine" and put the issue to rest as a one-off incident, rather than dig deeper within the organization. The speed of intervention and prompt resolution leaves the oversight question virtually unresolved.

Question #4. Which metrics/tools do you use to quantify reputational risk exposure?

Finding #4. The reported metrics varied by geographic focus of the business, customer segment (retail or government) and whether environmental or social features were primary elements of the service or product offered by the company.

Question #5. Who regulates reputational risk currently? How is it likely to evolve?
Finding #5. Respondents indicated either their sector-specific financial or consumer conduct authority, with a small percentage commenting on the politically driven nature of lobbying activities and conflicts of interest.

I conducted the same survey concurrently among participants, citing climate risk and the adequacy of environmental policies as an example of emerging reputational risk. Although the group discussions were not officially recorded within the survey, when addressing how individual business functions may be affected in terms of accountability for climate risk, the participants exhibited a wide disconnect. As global government regulations affecting business disclosures of climate risks are in flux, I am looking to redefine this disconnection through further research.

The participants' feedback confirms that built-in transparency incentives are a way to address the inevitable reluctance of an organization to change. They stir cross-sectional learning, remove the limiting barriers of a "cost of doing business" mindset and potentially open the door to sustainable innovation from within. Yet, having acted as a facilitator in group settings for a number of years, I am fully aware that the reluctance to provide feedback is more easily overcome when input is rendered anonymously or when it is not easily attributable (e.g., in the large forums I have described for my own surveys). Experimenting with game-like environments can be an effective way to increase engagement and make the process more rewarding for participants.

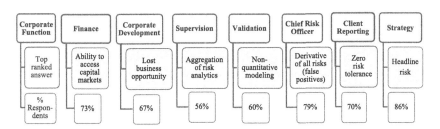

FIGURE 8.1 Survey: Managing reputational risk. **Survey question**: *"How is reputational risk defined across corporate functions?"* Note: Percentage of respondents supporting the highest voted definition of reputational risk by functional expertise/corporate function. Author's research as presented at the 2019 GRC Summit. Survey reflects aggregate results from 2015 to 2019 field study from 1,000 participants, 90%+ filling rate in all survey responses and less than 10% needing further clarification on more than three questions.

GAMIFICATION AND ENVIRONMENTAL SUSTAINABILITY

When discussing environmental sustainability and issues surrounding biodiversity, or when defining the boundaries of inclusiveness in group settings, it is hard to establish baseline knowledge. In turn, this makes it difficult to start the dialogue on possible scenarios and action plans. In fact, what makes scenario thinking more difficult is that it deals with sustainability outcomes that not everyone (if anyone) has dealt with and can relate to. Gamification, the art and science of designing a game-like setting to learn interactively while solving real-world problems, may help.

In multidimensional scenarios like the ones needed to describe institutional levers affecting environmental damages created by climate change, a system approach provides a powerful tool. It can lay out an integrated assessment of a system's environmental readiness, whether that system is a community, a company, a country or the entire planet, and drive optimal planning for climate adaptation. For example, the climate solution simulator En-ROADS, created by the US think tank Climate Interactive and the MIT Sloan Sustainability Initiative, offers a premier example of the value of simulation models coupled with gamification for engaging a group on climate solutions. Moreover, in the case of En-ROADS, it helps address the balance between decisions that favor a low carbon transition and socio-economic metrics that directly affect sustainable development.

These tools make it possible to employ a system approach in a group setting that allows individuals to experiment with potential solutions using real-time games. As a practitioner, I have experienced first-hand the value of this approach. It has a much better chance of encouraging discussions about climate and environmental sustainability that are based not on conflicting viewpoints but on shared value and a shared commitment. It brings a spectrum of problems affecting humankind to the human brain in a tangible, actionable form that can break down psychological barriers, anxieties and fears.

HOW TO BUILD A GAMIFIED ENTERPRISE ENVIRONMENT?

If we have done our job right, the business context in which we operate has recognized and embraced some degree of transparency in fostering both internal and external engagement on its sustainability roadmap. The next step is encouraging the informal organization to talk in a way that fosters positive behavioral changes and promotes day-to-day decision-making that is aligned with the long-term environmental or societal commitments set by the organization. The purpose of building a *gamified enterprise process* is to elevate group-level behaviors and actions beyond individual projects – no matter how impactful – into scalable decisions that can sustain the momentum and buy-in.

An example is the SaaS platform WeSpire. WeSpire has pioneered the use of games and other interactive technologies to deliver digital sustainability engagement programs at Fortune 500 companies that want to engage and motivate their employees. Built on the behavioral insights on the science of persuasion established by Prof. Robert Cialdini, the work of WeSpire aims at improving employee interest and participation in corporate sustainability initiatives to help meet corporate commitments. The followers of the Stanford Persuasive Technology Lab will recognize the incentives set forth by employing behavior design models that bring visibility and positive feedback to the forefront of employee engagement in the WeSpire programs.

Now let us broaden the lens to the web of daily decisions made by the *informal organization outside of sustainability initiatives*. I believe that building gamification in an enterprise context to help promote specific corporate commitments at the functional level is likely to empower the informal organization to actively participate in setting sustainability objectives. Further, solutions will be translated into shared achievements.

The case of introducing a water sustainability roadmap by leveraging the design of Collective Awareness Platforms (CAPs) offers a valuable application to start from. CAPs are web platforms that bring together social network theory and the power of collaborative actions. In their contribution to the January 2020 edition of the journal *Sustainability*, Ksenia Koroleva and Jasminko Novak describe both the theoretical background and existing applications of gamified incentives models to solve sustainability challenges. These models motivate participants to actively participate in sustainability domains and adopt both an individual and a community lens.

Through CAPs, participants undertake a series of activities that engage them in a user-targeted way (in order to avoid bias factors such as gender). A series of built-in incentives, such as points systems or badge awards, encourage the user to stay engaged. An example could be a system that motivates participants to save energy or reduce water consumption over a period of time. The system rewards individual participants, but the way it engages them is within the context of their local community needs (Figure 8.2). The intuition behind the model presented by Koroleva and Novak is quite novel. It draws together both the user motivations and contextual elements that may hamper an individual's desire to help build awareness, for example, of water conservation in their community or plan for water scarcity.

An increasing number of direct applications of a similar CAP design can be found in the areas of sustainable mobility, food waste and recycling, and regenerative agriculture.

Author's Note: Most of the exercises I have conducted over the years through hybrid learning models have been experimental. They rely on software platforms to define and record user engagement, plus live group discussions to reinforce learning and share feedback. Therefore, they have given more weight to "individual awareness" or "community impact" rather than defining user-level incentives to participate and maximize both personal and group rewards. The value of what I call "gamification for good" has clearly evolved into formalizing accountability mechanisms

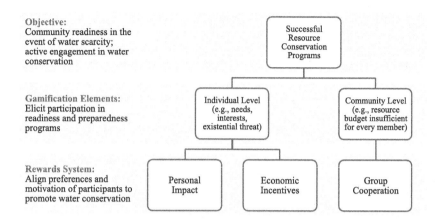

FIGURE 8.2   Gamification and water sustainability model. Author's intuition on the application of Koroleva and Novak's research.

across all levels of a group setting (an enterprise, for example). This goes beyond the prescriptive design that the *formal* organization, with its web of corporate functions and straitjacket of role responsibilities, may have tightly defined, and it therefore builds inclusive and creative networks.

MANAGING NON-FINANCIAL RISKS

Mapping the non-financial risks that a business is facing is a key step in setting its sustainability commitments. Gamification is a step past just making the business or the investment case for sustainability. Thinking of environmental, social, governance and non-financial risks that have deep financial consequences if left unattended may help balance shorter term trade-offs for an organization.

In the summer of 2019, the US association Business Roundtable issued the "Statement on Purpose of a Corporation," an open letter that was signed by over one-third of Fortune 500 CEOs. According to Bill McNab, former CEO of Vanguard, focus on long-term value creation is indeed about "better serving everyone: investors, employees, communities, suppliers and customers" (Business Roundtable 2019). The statement was issued at a time of significant societal challenges affecting delivery and redistribution of economic value. Since then, public and private companies have been increasingly rethinking their multistakeholder engagement as the investment community continues to promote more thoughtful disclosure of environmental, social and governance factors.

How should organizations address non-financial uncertainty in a purpose-oriented society? There are three key factors to consider:

#1 Non-financial uncertainty yields financial consequences
Investors and third-party providers that evaluate credit risk are increasingly seeking to incorporate ESG considerations that are more transparent and

industry-specific. For example, according to credit rating agency Standard and Poor's (S&P), approximately 10% of corporate rating assessments released by its global team over the two years to the end of June 2017 were affected by environmental and climate-related concerns. Most recently, the lack of oversight of ESG practices has triggered instances of financial distress among public utility companies and major international conglomerates – many of which occurred after the S&P study.

The recent launch of the listed financial derivative products featuring sustainability performance resets in favor of socially and environmentally conscious underwriters reminds risk practitioners, once again, of how financial innovation enables capital markets to assign a "fair value" to financially material risks, including those generated by non-traditional sources.

#2 Monitoring of reputational risk requires active management of ESG concerns

Most frequently, the magnitude of the reputational enterprise value that is at risk reflects the degree of alignment between corporate commitments and corporate behaviors. It calls for an institutionalized channel of knowledge transfer within an organization. From gender-based pay imbalances to the ratio of management to median employee wages and degree of inclusiveness of vendor policies, corporate commitments are increasingly traceable. Yet they remain less tractable by traditional business functions if addressed in silos. Activist stakeholders are pushing the boundaries of transparency and raising the bar for standardized disclosures of non-financial items. A most alarming outcome is the fact that they are also driving causality arguments on the longevity of corporate structures that are not adjusting fast enough to change.

#3 Regulatory readiness calls for outcome-focused innovations

Regulators continue to step up their game on non-financial risks (NFR) globally. In 2017, the Task Force for Climate-Related Financial Disclosures (TCFD) of the Financial Stability Board launched climate risk disclosure guidelines. These introduced a unique solution to an already highly regulated global marketplace by marrying compliance needs for NFR-tailored reporting with financial assessment of enterprise value at risk. In fact, by breaking up the guidance for implementation that accompanies the TCFD recommendations into four pillars (Governance, Strategy, Risk Management and Metrics/Targets), it also provides a roadmap to identify where purpose-oriented outcomes may intersect with financial value. In practice, sector-specific standards that leverage financial materiality, such as those offered by the Sustainability Accounting Standards Board (SASB), define a set of key reference metrics to measure corporate commitments against NFR-aligned targets – an essential step for market-oriented financial innovations to unlock the value of NFR mapping and reporting.

LOOKING AHEAD WITH PURPOSE

Although focused on climate risk, the TCFD recommendations provide a framework for comparing best practices and measuring an organization's NFR exposure. Outside of the realm of climate change, the four pillars of the TCFD guidance may become a valuable resource to map all the other non-financial items that affect existing operational work-streams – from procurement and vendor selection to service delivery and customer support – to direct sources of value. Active management of NFR calls for reporting performance against non-financial targets. Establishing a holistic process for doing so that cuts across corporate silos may become everyone's responsibility in a world that seeks purpose. Interweaving participatory learning in the form of gamification exercises can be a valuable first step in driving broader NFR literacy and pushing organizational incentives in the direction of inclusivity and accountability.

As the universe of reputational risks continues to expand, deploying a programmatic approach to the implementation of impact-oriented goals requires businesses to address group-level behaviors by promoting collective participation to sustainability initiatives. Gamification and Collective Awareness Platforms are emerging as impactful organizational tools to embed a component of participative learning and purpose in everyday interactions in the workplace while building new metrics and accounting for progress.

Can organizations address controversies related to human rights in a similar fashion?

How can businesses harness data to close the gap on the social dimensions of ESG?

THE LEARNING JOURNEY – THE INFORMAL ORGANIZATION

Stirring group-level behaviors and workplace interactions in support of impact-oriented goals requires businesses to embrace a programmatic mindset and leverage analytical approaches to measuring progress. The following checklist highlights the stepping-stones for the reader to contribute and affect change from within:

• Define the pillars of the informal organization. Which are the relationships and mindsets that stir internal processes and people, and influence the delivery of business value every day?
• Which reputational risks affect the business or its ecosystem the most? How are they discussed internally? List the stakeholder groups that are associated with reputational risk areas for the business. List examples of business decisions that have resulted from stakeholder engagement on key reputational risk areas. What worked? What did not?

- By referencing the SASB Materiality Map® tool, identify a set of metrics that are most representative of these reputational risks.
- Is gamification actively utilized within the business? Leverage web-based applications such as En-ROADS to gauge how the organization is most likely to be affected by environmental policies. How is the business positioned to address the short-term outcomes of those scenarios? If the business has released a TCFD framework, compare your En-ROADS scenarios with the reported information. Identify the areas where a higher degree of transparency will positively impact the organization (e.g., either as a strategic advantage or because of lower reputational risk).
- Drawing from your visualization experience in En-ROADS, organize a board game following the Play it Forward roadmap as laid out in the Technical Note. Which team members would you invite as participants? How would their interaction be a source of sustainability value for the business?

TECHNICAL NOTE – PARTICIPATORY MODELING FOR GAME DEVELOPMENT

Game development for use in business settings is both an art and a science. The most successful platforms usually start as board games. They then leverage the interaction of test groups and feedback among cohorts to implement them digitally and scale their use to reach larger audiences. In many cases, the game may become an attractive tool for multi-stakeholder engagement outside of the organization. What all platforms have in common is that they employ advanced methods of participatory modeling to render the social settings in which business decisions are made. Participants interact with one another, and interest groups are formed or entirely missed.

The example presented in this section follows the development of the Play it Forward (Dewulf, 2010) game, which encompasses a board game environment and a digital implementation. The goal of the exercise is to create sustainable outcomes in the design of product and process innovation in businesses. The participants are asked to wear a "People, Planet and Profit" hat when faced with potential challenges and opportunities in the context of product or business model innovation. Two groups are formed. One represents the big picture – market dynamics and long-term forces of sustainability at play in the business ecosystem. The other group is asked to maximize the value of technological and operational advances needed to adjust to those emerging trends and position the company to capture the upside. Each group is tasked to find the most profitable and sustainable business solution while influencing their colleagues and other stakeholders to join them on the journey (Figure 8.3).

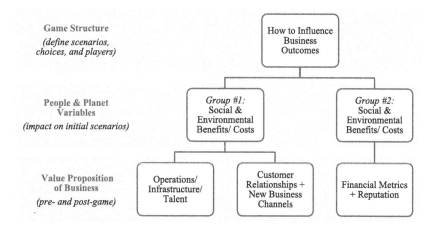

FIGURE 8.3 Gamification pillars for sustainable market and product design. Author's intuition on the set-up of the game Play it Forward. Players in Group #1 would swap team with Group #2 in a repeated game setting. In the first game, Group #1 will be assessing a series of environmental costs and benefits to influence the way the business delivers value through its operations, infrastructure, talent or its customers (existing and prospective). Group #2 will focus on a set of environmental benefits and costs that drive company profits and/or reputation higher. As the game evolves and the groups propose their suggested roadmap for the business to improve People, Planet and Profit returns in light of new sets of operating scenarios presented at inception of the game. Collectively the groups arrive at an optimal set of decisions that would make the business resilient in the Play it Forward model – either through new market, process or product development – regardless of the initial conditions of the game.

THE IMPORTANCE OF ROLE PLAY

In a computer-supported environment, the trade-offs in group versus individual objectives become highly visible. The need to selectively choose which trade-offs to tackle as a group and what to prioritize as an individual are key competencies that the game setting allows participants to develop.

A compelling example of the importance of role play in the development of cooperative models of action can be found in the *River Basin Game* by the University of Twente in the Netherlands. Systems like water management decisions for product and process innovation in hydrology are highly complex. For participants to clearly understand the role played by each decision, they need to be equipped with an integrated view of the challenge at hand that is rendered from the perspective of each group operating within the system.

By providing a transparent environment for participants to interact, building cooperative networks to advance sustainable and profitable business solutions becomes a matter of choice rather than a matter of specialized knowledge or individual belief. The introduction of metrics of socio-economic development and environmental protection, such as the ones commonly utilized in the UN

Sustainable Development Goals (SDGs), is invaluable in role plays that switch the goal of maximizing financial performance while complying with UN SDG metrics from one group to the other. A sample of development indicators presented in the River Basin Game includes, among others, indicators of food supply and water availability in the region, as well of dependence on natural resources not available locally.

WHAT CRITICS MAY SAY

Virtual or physical domains that host collaborative environments, such as the Play it Forward or the River Basin Game, rely on a set of scenarios and contextual variables that may hinder progress in decision-making for the group or even individual participants. In board game versions, such variables are introduced with the use of cards representing a potential, one-off external factor that may impact the evolution of product or business innovation. It could be rendered in the form of an emerging trend or a new and valuable piece of information that enhances the ability of the group to arrive at a consensus view of progress in terms of a societal or financial benefit or cost. Nevertheless, it is represented as a one-off situation or extreme event that the group is asked to adapt to.

By broadening the range of feedback that participants can provide in the form of an evaluation of impact over time, as opposed to a one-off event, this feature could in fact be viewed as an opportunity to reflect on the trajectory of improvement (or lack thereof) introduced to the system. One-time events that skew the balance between societal and financial benefit or cost over the life of the game can be introduced as an opportunity to define what a path of improvement and incremental change may look like in the existing organizational setting.

REFERENCES

Business Roundtable. 2019. "Statement on the Purpose of a Corporation." https://s3.amazonaws.com/brt.org/BRT-StatementonthePurposeofaCorporationJuly2021.pdf.

Cardoso, B., M. Ribeiro, C. Prandi, and N. Nunes. 2019. "When Gamification Meets Sustainability: A Pervasive Approach to Foster Sustainable Mobility in Madeira." *Proceedings of the 1st ACM Workshop on Emerging Smart Technologies and Infrastructures for Smart Mobility and Sustainability* 3–8. doi:10.1145/3349622.3355449.

Dewulf, K. R. 2010. "Play it Forward: A Game-based Tool for Sustainable Product and Business Model Innovation in the Fuzzy Front End." *Knowledge Collaboration & Learning for Sustainable Innovation.* ERSCP-EMSU Conference, 25–29 October, 2010. Delft, the Netherlands. http://resolver.tudelft.nl/uuid:eb7556af-8f93-4a3a-826b-e8ed6de72aa1.

Eker, S. 2021. "Testing the Economic Impact of Climate Change in En-ROADS." *Climate Interactive Blog,* 4 March, 2021. https://www.climateinteractive.org/analysis/economic-impact-of-climate-change-in-en-roads/.

Falsarone, A. 2019. "Seeking Purpose: Active Management of Non-Financial Risk." *Center for Financial Professionals* issue 13. https://www.cefpro.com/wp-content/uploads/2019/11/Alessia-Falsarone-article.pdf.

Financial Stability Board. 2017. "Recommendations of the Taskforce for Climate-Related Financial Disclosures." Final Report. https://www.fsb-tcfd.org/.

Global Footprint Network National Footprint and Biocapacity Accounts. 2021 edition. Downloaded 2 May, 2021 from https://data.footprintnetwork.org.

Hoekstra, A. Y. 2012. "Computer-Supported Games and Role Plays in Teaching Water Management." *Hydrology and Earth System Sciences Discussions* 9: 1859–1884. doi:10.5194/hessd-9-1859-2012.

Huber, M. Z., and L. M. Hilty. 2015. "Gamification and Sustainable Consumption: Overcoming the Limitations of Persuasive Technologies." *ICT Innovations for Sustainability.* Cham: Springer, 367–385. doi:10.1007/978-3-319-09228-7_22.

Koroleva, K., and J. Novak. 2020. "How to Engage with Sustainability Issues We Rarely Experience? A Gamification Model for Collective Awareness Platforms in Water-Related Sustainability." *Sustainability* 12 (2): 712. doi:10.3390/su12020712.

Nordby, A., K. Øygardslia, U. Sverdrup, and H. Sverdrup. 2016. "The Art of Gamification; Teaching Sustainability and System Thinking by Pervasive Game Development." *The Electronic Journal of e-Learning* 14 (3): 152–168.

Nyström, T. 2017. "Gamification of Persuasive Systems for Sustainability." *Sustainable Internet and ICT for Sustainability:* 1–3. doi:10.23919/SustainIT.2017.8379815.

Stevens, S. H. 2013. "How Gamification and Behavior Science Can Drive Social Change One Employee at a Time." In *Design, User Experience, and Usability. Health, Learning, Playing, Cultural, and Cross-Cultural User Experience,* DUXU 2013, Lecture Notes in Computer Science 8013, edited by A. Marcus, 597–601. Springer, Berlin, Heidelberg. doi:10.1007/978-3-642-39241-2_65.

Sustainability Accounting Standards Board. 2018. SASB Materiality Map®. https://materiality.sasb.org/.

9 The Impact of Data on Human Rights

The field of human rights touches myriad disciplines: from public health to international law, forensic studies, community relations and public policy. It falls in the realm of "social science measurement." Over time it has usually been associated with the work of advocacy groups, NGOs and multilateral organizations, who seek to tackle violations of human rights in countries located in conflict-prone regions, or that are at risk because of their role in global supply chains.

It is worthwhile noting that this chapter will not address the impact of treaties, governmental policy action on human rights, or the linkages with political instability and country risk (unless relevant for estimating vulnerabilities in international commerce hubs). Rather, the discussion will focus on how businesses can begin to leverage data in unveiling and managing vulnerabilities in their ecosystem, within and beyond supply chains.

In 1992, the University of Pennsylvania Press published the landmark collection of studies edited by Thomas Jabine and Richard Claude titled *Human Rights and Statistics: Getting the Record Straight*. To date, it is by far the strongest reference for anyone interested in collecting human rights data and building credible impact-oriented metrics.

In 2015, the United Nations General Assembly ratified the 2030 Sustainable Development Agenda, and in response, the UN Sustainable Development Goals (SDGs) were identified – a set of core variables required to achieve the targets set out in the Agenda. Private sector participants, whether businesses or investors, have been called to take stock of their "human rights blueprint." They must investigate whether the business practices and lobbying activities they employ can be aligned with measurable outcomes toward achieving the SDGs. This must go beyond minimum compliance with international standards or avoidance of human rights abuses. To get there, a series of questions must be answered:

- What human rights data can be collected in a credible and statistically significant manner?
- Is the data currently available meaningful and helpful to organizations in their roadmap to fulfill their social sustainability commitments?
- Most importantly, who is the human rights data currently in use built for?

DOI: 10.1201/9781003212225-10

NORMS, DATA AND ACCOUNTABILITY

The release of the UN Guiding Principles on Business and Human Rights (the Principles) in 2011 marked the launch of the first official global standard for businesses to prevent, address and remedy adverse impacts associated with their activities, whether directly or through procurement and partnership channels globally. They also marked the beginning of a decade-plus of notable attempts by government agencies and international organizations to encourage outright business integration of human rights-sensitive and transparency-focused operational efforts.

Inevitably, what I call the "decade of norms" has been followed by a nearly concurrent "decade of data transformation." Academics and public and private sector practitioners have sought to quantify, on one hand, the effectiveness of norms and government policies. On the other hand, they have sought to gauge how vulnerable business conduct and operations are to violations of human rights principles. Definitions, metrics and frameworks have emerged to help companies define the subset of human rights that are most at risk in their operations and business dealings and attempt to minimize the harm associated with those priority areas in a dynamic, rather than prescriptive, way.

Mandatory human rights disclosure regulations have been introduced across OECD countries. This has led businesses to shift toward adopting impact assessment frameworks that represent their own social and human capital strategy and reflect their operational footprint and their ecosystem of suppliers, customers and partners. In turn, this has led to the emergence of knowledge platforms that aid data collection and visualization of a variety of metrics.

The changing nature of human rights metrics reflects this move from normative and academic definitions to actionable impact frameworks: From the narrow scope of cataloging human rights violations, to providing a snapshot of overall human rights conditions in a region or sector (such as working conditions in a supply chain or in conflict-prone areas), to measuring the effectiveness of governments in upholding human rights and protecting populations at risk.

As business accountability has become tightly connected with impact assessments, organizational vulnerabilities linked with human rights are increasingly seen by investors and the general public as proxies for the effectiveness of corporate decisions in those areas. The quantification of impact of human rights policies plays a key role in that journey, from the types of data that are most valuable for a business to collect, to frequency of evaluation and functional representation and accountability by integrating human rights compliance with a risk mitigation mindset. Companies are also expected to seek opportunities to create positive impact by promoting policy dialogue around social wellbeing beyond the narrow focus of their operations or geographic markets.

HUMAN RIGHTS AND CONTROVERSIAL BUSINESS ACTIVITIES IN PRACTICE

Historically, international norms and legal frameworks have defined the boundaries of human rights by drawing attention to violations of ethical and social conditions.

The measurements that have followed reflect those attempts to define violations and lead to interpretations that reflect standards of local living conditions and cultural boundaries. Practitioners and scholars have contributed to a significant body of work for the purpose of monitoring and mitigating noncompliance.

Practical approaches to human rights risks stem from several tools, from event-based metrics (mainly for forecasting purposes) to survey-based metrics (which focus on the evolution of a target population at risk). Table 9.1 highlights the most comprehensive categorizations of human rights statistics, including the emerging use cases for each of the following:

- **Event-based statistics**: Help define the context in a descriptive way, by providing the facts behind human rights abuses or noncompliance with international norms. A notable example is real-time tracking of media outlets through open-source automated systems (also known as Events Media Monitoring, or EMM). This is usually guided by a dictionary of terms to code events for data extraction. EMM requires routine verification to ensure selection or significant biases are accounted for.
- **Standards-based statistics**: These help identify the scale or range of acceptable situations and flag what is unacceptable. The Cingranelli and Richards (CIRI) database, among others, provides historical context for a reference set of standards-based statistics spanning over 200 countries. Practitioners who assess the SDG impact on economic development will find the indicators on Women's Economic Rights, along with CIRI's Empowerment Rights Index, which are valuable statistics for benchmarking the effectiveness of their policies (from responsible procurement to country risk assessment) and the effectiveness of their stakeholder engagement initiatives.
- **Survey-based statistics**: Unlike event-based measures, surveys allow for generalized testing through random samples of a target population at risk. The rigor employed in both designing the survey (a selection of well-defined answers) and identifying population samples ensures that impact assessments capture the evolution of the target population. Patterns in responses help define margins of error.
- **Socio-economic statistics**: Built from the official statistical databases of sovereign entities and subnational governments, this category represents a broad set of metrics that define the effectiveness of governmental policies with regard to socio-economic outcomes. Open access to these data in disaggregate format allows practitioners to select relevant country-level measures and analyze the degree of inclusiveness and/or equality in benchmarking studies (e.g., as a baseline to trends that may have emerged in localized surveys). A notable shortfall of data sourced from Official Country Statistics is that access to sources and quality of reporting may remain a challenge for those jurisdictions that lack policies, infrastructure and technical assistance to support a comprehensive data collection effort.

TABLE 9.1
Human Rights Measures

Category	Context	Reference Metrics	Emerging Use Cases
Event-based	*What happened, when, and to whom*	Number and Frequency of Incidents (Quantitative)	• Forecasting purpose (likelihood of events)
Standards-based	*How reported violations compare to thresholds*	Qualitative Assessments (e.g., more, less, average)	• Comparability assessment (effectiveness of policies and stakeholder engagement)
Survey-based	*Generalized testing of at-risk population*	Structured or Semi-Structured Sampling	• Evolution of target population at risk (impact assessment updates)
Socio-economic	*Reference of national statistics as proxy measures*	As reported by National Statistical Agencies	• Evolution of target population at risk; cross-country studies (impact assessment updates)

Source: Categories listed follow the definitions by Landman and Carvalho (2010). Reference metrics and use cases drawn from the author's applied research.

The sample assessment in Table 9.1A and 9.1B helps us to apply a multistakeholder lens to human rights measures when analyzing the impact of controversial business practices, such as involuntary resettlement of indigenous populations.

THE DILEMMA OF MEASURING HUMAN RIGHTS EXPLAINED

The dilemma of quantifying human rights starts with the "why."

What we are measuring when addressing human rights very much depends on why a business is focusing on the topic. Is it driven by compliance with international norms and laws? Is it due to investors' pressure for higher transparency? These have both been key drivers, with investors focusing predominantly on indirect impact to a business through its procurement cycle and its supply chain due diligence. For compliance purposes, event-based analyses and, increasingly, standards-based measures have been taking center stage for businesses. Web-based and AI-delivered supply chain reporting is now in place at Fortune 500 companies. This makes it easier to build a baseline, by geographic region or sector, by benchmarking single incidences of noncompliance.

Nevertheless, the task of monitoring and evaluating business complacency toward human rights violations and the management of human rights risks prior to a violation occurring is still largely unresolved. We are living in a time when corporate sustainability commitments are met with the rollout of ambitious global goals and responsible procurement policies. Yet few organizations have truly taken the time to establish a policy dialogue on human rights compliance as a way to build a mindset of organizational advocacy for populations at risk.

TABLE 9.1A
Analysis of Controversial Business Activities

Controversy	Stakeholders	Human Rights Lens	Methodology
Forced Resettlement Due to Land Acquisition *Physical displacement of low-income ethnic communities due to infrastructure project development* Regional Focus: **Latin America** Population at risk: **Indigenous Peoples**	• Indigenous Populations • Project Implementers • Local authorities • Financial institutions	• Involuntary resettlement (eviction) • Economic displacement due to access restriction to land and resources • Mitigation and remediation measures (culturally appropriate) **Socio-Economic Profile:** • Local renewable energy market in high demand for infrastructure development projects • Weak land conservation and biodiversity laws	**Impact Assessment:** • Historical precedents of resettlement practice • Claims of at-risk populations lack representation • Kept uninformed; consultations with local community by project developers lack timelines and transparency on land reuse • No biodiversity legal obligation to respect national heritage lands, nor to remediate controversial incidents • Occurrences of negotiated and agreed settlements (including expropriation) lack monitoring and evaluation of culturally appropriate, non-discriminatory practices

Source: Author's intuition and research. For illustrative purposes only.

However, even if the monitoring and evaluation of deviations from human rights policies do not advance the internal and external policy dialogue, the work is still worthwhile. A conceptual framework, as opposed to a measurable and executable accountability statement, can still be produced. A compelling example of the dichotomy between near-term compliance and long-term policy dialogue is that of evaluating and monitoring the impact of modern slavery in a global, interconnected setting.

MODERN SLAVERY: THE ACCOUNTABILITY LENS

Modern slavery encompasses a variety of subcategorizations of potential human rights violations, from forced and child labor to human trafficking and exploitation. It is a phenomenon that continues to draw escalating international scrutiny and increasing efforts to legislate its definitions and punish its outcomes.

TABLE 9.1B
Social Impact Assessment – Land Acquisition Example

Purpose of Social Impact Assessment:	Step I. Social context:	Step II. Risk mapping:	Step III. Envisioning outcomes:
1. Define social context	• Social inclusion of indigenous people and environmental protection of cultural heritage sites at risk	• Safety standards and compliance with the rule of law	• Established social bonding on respect of local land rights
2. Identify sources of risk	• Local sourcing of products and services	• Community priorities and stakeholder engagement	• Maximization of value and respect of cultural heritage sites
3. Describe potential outcomes (adverse and non) and their timeline	• Respect of laws	• Social license to operate and business performance	• Lower level of inequalities within local communities and workers' populations
	Step IV. Mitigating/magnifying factors:	**Step V. Negotiation/mitigation goals and timeline:**	**Step VI. Opportunities for impact monitoring:**
4. List mitigating/ magnifying factors	• Social cost of migration/displacement support	• Project-level impact on community access to natural resources, infrastructure, essential services	• Fair land use improves relationships with local communities
5. Define goals and timeline for negotiation/ mitigation action	• Local talent pool (fair wages, skill-building programs)	• Timeline: site development activities, construction and transport activities	• Potential for job creation and community investment programs
6. Address opportunities for impact monitoring	• Perception of hazardous/ safe working/ living conditions	• Ethical and informed consent, negotiation and compensation (physical and economic displacement)	• Healthy and productive work environment, incl. safe mobility of workforce
			• Absence of violations in compliance with local laws with respect to shared use of land, healthy living conditions
Key priorities (identified from Social Impact Assessment):	**Social and human priorities:** Seamless transfer of land use without harm to local population and/or environmental degradation and potential for economic development	**Business priorities:** Timely completion of land transfer for infrastructure project in adherence with fair business conduct	

Source: Author's intuition and research. For illustrative purposes only. Reference methodology on impact valuation method: WBCSD (2019).

The global agenda for sustainable development set forth by the UN SDGs, and the commitments undertaken by leading private and public sector participants in the UN Global Compact Initiative, has brought attention back to the enablers of change in public and private accountability for modern slavery and deflected attention from the false precision of definitions.

As for all other violations of human rights, quantifying direct and indirect impacts of modern slavery on all parties involved continues to make the engagement of the private sector compelling beyond policymaking. The accountability lens for preventing modern slavery relies on a dynamic measurement framework. As with other categories of human rights, analyzing trends and parameters that may act as signals will most likely include event-based data collection, comparability studies based on standards and survey analytics. These tools will be used to examine how effective policies are and capture new sources of data, many of which are linked to digital interactions. Verification and validation of information sources – whether through aerial maps or witness interviews on the ground – will continue to face the critical challenges that de facto completeness of the evaluation presents. For example, are all the relevant data points being captured? Are reference standards applicable to the specific situation? Are potential biases fully understood by practitioners and academics?

A READINESS FRAMEWORK FOR BUSINESSES: VULNERABILITY RISK MAPPING

The classifications introduced in Figure 9.1 are applicable across human rights controversies, including modern slavery. A variety of data sources are involved in carrying out a substantive evaluation of business readiness and resilience to reputational and legal risks associated with modern slavery. Therefore, it is important to remember that issues such as data latency and the traceability of information as disclosed in the public domain affect the collection of key findings. This applies whether conducted retroactively or used to create forward-looking policies. Mindful of both latency and traceability of information (whether qualitative or quantitative), the compilation of a social impact assessment, as described in Table 9.1A and Table 9.1B, is a first step to building a readiness framework to counteract modern slavery, where accountability starts with an exhaustive and reliable data collection exercise.

After the introduction of the UK Modern Slavery Act in 2015, government-run platforms that aid transparency and corporate compliance, while also enabling the creation of digital repositories of modern slavery incidents, have quickly evolved. These important steps have contributed to speeding up the integration of social and human dimensions in business readiness planning for the purpose of regulatory compliance. However, the visualization of vulnerabilities and the quantification of negative impacts generated by businesses has not yet been met with an evaluation of the ways that organizations can help and improve human rights. Employing the framework set forth by the World Business Council for Sustainable Development in its Social and Human Capital Protocol is a necessary step for businesses to leverage vulnerability assessments to also define how they can create positive change in the realm of human rights. (See Technical Note for a discussion of reference data sources and mapping of modern slavery risk in the direct operations of a company or at the regional level through its supply chains.)

THE OECD DUE DILIGENCE GUIDANCE FOR RESPONSIBLE BUSINESS CONDUCT

Since its adoption in 2018, the OECD Due Diligence Guidance for Responsible Business Conduct has been the landmark reference for transnational companies that wish to avoid social and environmental harm – in the regular course of doing business, in their operations and supply chains, or when engaging in partnerships (Figure 9.1). To make it actionable, practitioners usually refer to the *due diligence* focus associated with the OECD as the "supply chain responsibility" of a business.

The guidance offers sector-specific references for agriculture, minerals, garment and footwear supply chains, along with practice papers for companies involved in extractives and financial sectors. It identifies six stages of the due diligence process that, when employed dynamically and reflected both in business policies and management systems, are designed to ensure transparency when early signs of potential harm caused by business actions are detected.

Before the guidance was released in 2018, the investor lens on human rights due diligence had focused primarily on measurability. It therefore drew heavily from either voluntary disclosures by companies or disclosures mandated by international laws on human rights. Examples of adverse impacts on human rights covered by the guidance include forced labor; wage discrimination (for equal work or work of equal value); gender-based violence; failure to engage with indigenous peoples impacted by business activities; restriction of people's access to natural resources such as clean water and reprisals against civil society and human rights defenders. Nevertheless, over the years, corporate efforts toward the collection of exhaustive evidence on alleged violations of international norms have failed to partner strongly with advocacy groups and local NGOs. But this is needed to prioritize engagement with key stakeholders and advance early dialogue on some of the most pressing human rights issues.

The launch of digital platforms has provided the much-needed infrastructure to help reconcile the frequency and severity of breaches with the facts reported by local groups. By aggregating material evidence on the human rights issues most at risk at the country, region, sector and project level, businesses are increasingly able to draw scalable and sustainable due diligence roadmaps across the management systems they operate.

As a sustainable investor, I have adopted a responsible due diligence framework to address the readiness of businesses at high social risk across a variety of countries and value chains. Because of that experience, I encourage using *a gender lens* in due diligence planning and execution. This is a direct way to deepen

FIGURE 9.1 Due diligence for impact. OECD (2018) and author's intuition.

the scope of the analyses when addressing business and societal impacts caused by human rights issues. This is particularly so in situations where women may be disproportionately at risk. Such situations include where women represent the majority of the local workforce; where their employment and living conditions affect the wellbeing and livelihood of their communities; or in geographies at high risk of conflict where women's vulnerabilities are exacerbated by their working status, a lower degree of literacy and, possibly, because they are indigenous. Applying the gender lens may be as simple as enabling women's participation in consultations and negotiations, specifically as related to adopting gender-sensitive mitigation and remediation practices.

New Types of Human Rights

The boundaries of what constitutes the human rights umbrella are rapidly evolving. In fact, scholars believe that the Fourth Industrial Revolution has made it harder to define where the boundaries should lie when it comes to monitoring potential human rights abuses and/or identifying and preventing accidental violations.

Consider these elements of the Fourth Industrial Revolution: digitalization and the future of smart cities and life-long learning through cloud-based applications; the movement for open data information, corporate stewardship efforts on data assets and the implications of privacy laws; distributed workforces with remote working and reduced human interaction. All these elements point to the potential for biases in algorithms and AI-focused innovation as unintended consequences of real-time monitoring. These may include the collection of identifiable information to determine at-risk categories of individuals as consumers, patients, or even employees, to list just a few.

Human rights laws, definitions and public perception will continue to evolve as a mirror of modern society, and data fields and information sources will inevitably reflect those changes. Metrics and data collection exercises will become valuable as historical points of comparability. By giving us a snapshot of a *point in time*, they will tell us whether there is "less or more" respect for human rights, "less or more" prevention of human rights abuses, "less or more" efficacy of government or company policies and tools. When an accountability framework for human rights readiness for businesses embeds the state of emerging vulnerabilities with their component of both social costs and potential social benefits, it will not become obsolete. The dynamic effort of mapping risk sources and embedding emerging trends into continued policy dialogue, and in-depth due diligence efforts will remain core as a proactive advocacy tool and a platform for multistakeholder engagement on human rights.

With corporate responsibility pressures mounting, digital platforms bring forth a decade of data transformation to help businesses close the transparency gap in the social context in which they operate.

Does the evolution of business models play a role in reframing sustainability risks into opportunities?

How could innovation ecosystems contribute to solve the impact challenge of businesses?

THE LEARNING JOURNEY – THE IMPACT OF DATA ON HUMAN RIGHTS

The evaluation of human rights vulnerabilities is foundational to the social impact assessment of businesses. The following roadmap guides the reader through the identification of gaps in responsible business conduct pertaining to human rights:

- Define the human rights dimensions that affect the sector and geographies where the organization operates. Choose first the controversy lens to define the boundaries of responsible business conduct. How have controversies evolved over time? Has the universe of stakeholders stayed the same or changed?
- Identify the populations at risk (e.g., event-based evaluation or due to the socio-economic profile of business activities). Have consultations or negotiations and agreed settlements taken place? Note potential discriminatory practices (either perceived or reported) that may have hindered mitigation of adverse impacts.
- Sketch a social impact assessment plan for the organization. Where can positive externalities be created? Who are the recipients of the social benefit the business is set to create? Consider business vulnerabilities from the lens of the social cost to the stakeholders at risk. Which

restrictions to human and socio-economic wellbeing pose the highest social cost? Can a mitigation roadmap be created?

• List data sources and potential indicators to begin mapping vulnerability risks. Refer to the Social and Human Capital Protocol by the World Business Council for Sustainable Development to identify negative and positive externalities (social costs and enablers of positive change). Note which roadblocks are unique to the indicators at hand that may challenge progress because of data gaps or reliability of underlying sources. If there were no roadblocks, how would such measures inform decision-making within the business? Are there any emerging risks that need industry-scale innovation to be fully researched and accounted for?

TECHNICAL NOTE – HUMAN RIGHTS DATA SOURCES

This section addresses the data gap that academic and nonacademic practitioners have experienced when approaching human rights issues. Most indices that comprise social dimensions are built bottom-up by aggregating granular indicators of individual risks. As a result, they may lack transparency and historical comparability by construction. In addition, as discussed earlier in the chapter, a gender-specific and stakeholder-focused lens should be applied when collecting data and evaluating outcomes, in order to assess an organization's broad social impact and its readiness to face future societal challenges. In fact, as many advocates of transparency in human rights initiatives continue to stress, incomplete datasets trace back to fragmented historical records. Practitioners and academics argue that those incomplete records may represent efforts to deliberately suppress any evidence of wrongdoing and complacency – in private and public sectors alike. Therefore, while it is necessary to employ a variety of data sources and methodologies as part of corporate due diligence efforts, it is not the ultimate solution (Table 9.2).

Modern Slavery: Vulnerability Risk Mapping and Modeling

The lack, comparability and quality of human rights data and the fragmentation of underlying sources of information remain roadblocks to developing a reliable impact assessment for businesses in their operations and across their supply chains. Computational science and artificial intelligence solutions are promising areas of research and development. For example, open digital maps can help to visualize the prevalence of modern slavery, and financial data can provide early warning of human trafficking through vulnerability modeling.

There is seminal work being carried out by the human rights knowledge platform Delta 8.7 and the United Nations University Center for Policy Research. This work models modern slavery as a country-level risk factor. It is an example of the prediction methodologies that combine data from a variety of sources to produce more reliable estimates of modern slavery. Leveraging survey-based statistics such as the Global Slavery Index by the Walk Free Foundation and

TABLE 9.2

A Selection of Global Human Rights Datasets

Database	Source Institution(s)	Human Rights: Categories and Scope	Methodologies
The Cingranelli-Richards (CIRI) human rights dataset https://www.humanrightsdata.com/	• CIRI Human Rights Education Initiative • U.S. Dept. of State Country Reports on Human Rights Practices • Annual Reports, Amnesty Intl. (USA)	• Standards-based • Government respect for 15 international variables incl. physical integrity rights, civil liberties, workers' rights, equality, women's issues	• Web-based aggregation • **Frequency:** Annual (1981–2011) • **Use:** Estimate effect of structural policy changes and humanitarian intervention • **Reliability:** Independent coding
World Bank ESG data portal https://datatopics.worldbank.org/esg/	• The World Bank and affiliated institutions (USA)	• Standards-based • Selection of 17 ESG themes at the country level. Six social themes include: education & skills; employment; demography; poverty & inequality; health & nutrition; access to services	• Proprietary sovereign data framework • **Frequency:** Annual (1997–2019) • **Use:** Country-level data visualization and dashboards that connect the sustainability of country's socio-economic and development needs • **Reliability:** World Bank platform
Human rights data analysis group HRDAG https://hrdag.org/	• Independent not-for-profit organization (Statisticians for Human Rights) sponsored by Community Partners (El Salvador; USA)	• Event-based • Data science and statistical support to carry out evidence-based assessment for social justice	• Proprietary evaluation framework • **Frequency:** Ad hoc, project-based • **Use:** Statistical analyses and expert evaluation of event-based data • **Reliability:** Project portfolio based on strict selection criteria

(Continued)

TABLE 9.2 (*Continued*)
A Selection of Global Human Rights Datasets

Database	Source Institution(s)	Human Rights: Categories and Scope	Methodologies
All Minorities At Risk Project (AMAR) https://www.mar.umd.edu/	• University-based research project, hosted by the Center for International Development and Conflict Management University of Maryland (USA)	• Socio-Economic • Country coverage of nearly 1,200 ethnic groups in countries with population over 500,000	• Proprietary dataset, tracks political, economic and cultural dimensions • **Frequency**: Annual (1988 - current) • **Use**: Comparative research of conflicts and ethnic mobilization; historical chronology • **Reliability**: University of Maryland web platform
Quality of Government (QoG) Institute https://www. gu.se/en/quality-government	• Independent research institute, Good Governance and QoG Studies. University of Gothenburg (Sweden)	• Survey-based (nearly 1,000 government, transparency and public administration experts) • Coverage of 100+ countries related to QoG policy variables on areas such as health, the environment, social policy and poverty (incl. sub-national data)	• Individual and aggregate datasets (incl. Eurostat) • **Frequency**: Annual (2004 - current) • **Use**: Comparative studies of quality of government; variable search and visualization tools in R and STATA; QoG indices and country-level indicators • **Reliability**: University of Gothenburg QoG researchers

(*Continued*)

TABLE 9.2 (*Continued*)
A Selection of Global Human Rights Datasets

Database	Source Institution(s)	Human Rights: Categories and Scope	Methodologies
Global slavery index walk free foundation https://www. globalslaveryindex.org	• G20 country-by-country ranking of populations affected by modern slavery and government response. Hosted by international NGO Walk Free (Australia)	• Survey (Gallup World polls) and event-based • G20 countries; national and regional maps and reports organized by prevalence/frequency, vulnerability scale; government responses; product-level risk mapping for supply chains	• Global index and disaggregated data • **Frequency:** Ad hoc (2019) • **Use:** Prevalence and vulnerability measures, assessment of government responses, mapping of at-risk product chains (G20), industry-level risk factors (e.g. fishing industry) • **Reliability:** Global Freedom Network; partnership with International Labor Organization and International Organization for Migration (IOM)
United Nations Human Rights Office of the High Commissioner (OHCHR) https://www.ohchr.org/ https://tbinternet.ohchr.org/	• Global repository of human rights treaties by the United Nations (USA and EU)	• Standards-based • Legal and statistical compilation of treaties within the UN system (database by country)	• Independent assessments of implementation of core international human rights treaties • **Frequency:** Annual (2010 – current) • **Use:** Monitoring of compliance with international treaties; country reports and implementation guides • **Reliability:** UN system (UN member states)
Human Rights Watch (HRW) https://www.hrw.org/	• International nongovernmental organization (USA)	• Norms-based (international human rights and humanitarian law) • World Reports (100+ country coverage of at-risk populations)	• Event- and interview-based country reports • **Frequency:** Annual (1989 - current) • **Use:** Thematic research of violations, education and advocacy on human rights • **Reliability:** Nobel Peace Prize awardee

(*Continued*)

TABLE 9.2 (*Continued*)
A Selection of Global Human Rights Datasets

Database	Source Institution(s)	Human Rights: Categories and Scope	Methodologies
Delta 8.7 https://delta87.org/	• Global knowledge platform of the UN University Centre for Policy Research and Alliance 8.7 (International coordinating groups)	• Norms-based and socio-economic development indicators for 193 countries committed to measuring impact towards UN SDG Goal 8.7 ("Take immediate and effective measures to eradicate forced labor, end modern slavery and human trafficking and secure the prohibition and elimination of the worst forms of child labor, including recruitment and use of child soldiers, and by 2025 end child labor in all its forms")	• Event- and expert interview-based global and regional dashboards • **Frequency**: Annual (current) • **Use**: Data visualization and mapping tools for monitoring frequency of human rights violations, government efforts and social protection; vulnerability mapping • **Reliability**: 35 experts on issues of modern slavery, forced labor, child labor and human trafficking
Amnesty International https://www.amnesty.org/en/	• International nongovernmental advocacy and research organization focused on human rights (UK)	• Survey and norms-based (159 individual country profiles, 5 regional profiles and a global report) • Follows human rights sub-categorizations in alignment with the Universal Declaration of Human Rights	• Event- and expert interview-based qualitative reviews • **Frequency**: Annual (current) • **Use**: Monitoring of trends and potential cause of current and future human rights violations; populations and regions at risk • **Reliability**: Global network of advocates, researchers and campaign supporters

Source: Author's research and open data sources as listed. Data sources are available in open access unless otherwise indicated. For illustrative purposes only.

the dedicated modern slavery module of the Gallup World Poll, the Delta 8.7 researchers are able to tailor the screening questions from survey respondents to clearly define select categories of the much larger universe of modern slavery cases, such as forced labor and forced marriage. The estimation of country-level vulnerability scores is carried out through Bayesian models. The goal of mapping modern slavery risk by country is focused on analyzing prevalence (e.g., where the probability of finding positive instances of modern slavery is higher). But cross-country comparability is likely to be improved when demographic and socio-economic development considerations are overlayed to address existing or future policy actions to lower country-level vulnerability.

SOCIAL FOOTPRINT: A MULTIREGIONAL INPUT-OUTPUT APPROACH

Human rights issues are set against a backdrop of multiple socio-economic and geopolitical dimensions and vulnerability assessments of businesses must be considered within that context. This suggests that aggregation of regional and sector-level indicators is best tracked through an *input-output accounting approach* to the operational and financial activities that create either a negative or a positive impact in terms of human and social capital. Similar to product life-cycle analyses for environmental dimensions, adopting an input-output accounting mindset when evaluating indicators of human and societal impact requires disaggregating them in the underlying social costs incurred by a business and/or the social benefits contributed by a business.

The reference analysis by researchers Hardadi and Pizzol on assessing labor-related impacts offers a methodological example on deriving metrics of human productivity (as either social benefits or costs associated with economic development) and metrics of human health and wellbeing (as either social benefits or costs associated with working conditions). By disaggregating indicators of regional economic development and working conditions, such as average occupational accident rates and average salary, Hardadi and Pizzol characterize a damage to employee health as having direct repercussions on the life-years of an individual (in fatality rates and life expectancy estimates), while a damage on human productivity is translated in a direct reduction of labor and economic production in terms of salary and employment or unemployment rates.

REFERENCES

Cingranelli, D. L., and D. L. Richards. 2010. "The Cingranelli and Richards (CIRI) Human Rights Data Project." *Human Rights Quarterly* 32 (2): 401–424. JSTOR. www.jstor.org/stable/40783984 Accessed 11 May 2021.

Clark, A., and K. Sikkink. 2013. "Information Effects and Human Rights Data: Is the Good News About Increased Human Rights Information Bad News for Human Rights Measures?" *Human Rights Quarterly* 35 (3): 539–568. http://www.jstor.org/stable/24518073 Accessed 11 May 2021.

Diego-Rosell, P., and J. Joudo Larsen. 2018. *Modelling the Risk of Modern Slavery.* July 17, 2018. Available at SSRN: https://ssrn.com/abstract=3215368 or http://dx.doi.org/10.2139/ssrn.3215368.

Hardadi, G. and M. Pizzol. 2017. "Extending the Multiregional Input-Output Framework to Labor-Related Impacts: A Proof of Concept." *Journal of Industrial Ecology.* doi:10.1111/jiec.12588.

Jabine, T. B., and R. P. Claude. 2016. *Human Rights and Statistics.* Philadelphia, PA: University of Pennsylvania Press. doi:10.9783/9781512802863.

Landman, T. 2020. "Measuring Modern Slavery: Law, Human Rights, and New Forms of Data." *Human Rights Quarterly* 42 (2): 303–331. Project MUSE. doi:10.1353/hrq.2020.0019.

Landman, T., and E. Carvalho. 2010. *Measuring Human Rights.* London: Routledge.

OECD. 2018. "OECD Due Diligence Guidance for Responsible Business Conduct."

Ruggie, J. G., and J. F. Sherman III. 2017. "The Concept of 'Due Diligence' in the UN Guiding Principles on Business and Human Rights: A Reply to Jonathan Bonnitcha and Robert McCorquodale." *European Journal of International Law* 28 (3): 921–928. doi:10.1093/ejil/chx047.

Soh, C., and D. Connolly. 2021. "New Frontiers of Profit and Risk: The Fourth Industrial Revolution's Impact on Business and Human Rights." *New Political Economy* 26 (1): 168–185. doi:10.1080/13563467.2020.1723514.

Vionnet, S., D. Friot, S. Haut, and R. Adhikari. 2021. "Screening for Human Rights Impact in Corporate Supply Chains – A Methodological Proposal for Quantitative Assessment and Valuation – Novartis Case Study." Working paper. Valuing Nature.

World Business Council for Sustainable Development (WBCSD). 2019. "Human and Social Capital Protocol." https://capitalscoalition.org/capitals-approach/natural-capital-protocol/?fwp_filter_tabs=guide_supplement.

10 The Next Decade
Innovation to Impact

The impact challenge faced by businesses entering the decade 2020–2030 is likely to be defined by three core areas for targeted action among private sector participants.

First, *the increasing importance of the Sustainable Development Goals and the longer term commitments by governments and public sector institutions to restore natural ecosystems and significantly rein in global warming.* The New Climate Economics Index, by the Swiss Re Institute, suggests that the transition to a carbon neutral society is expected to impact 90% of the world economy. That is according to statistics released in April 2021. The indirect effects of widening social divides, deteriorating natural capital and potential migrations are threats to the stability of all human life. Moreover, the global health crisis of 2020–2021 pushed the boundaries of what adverse societal impacts can become and the rapid outcomes they generate.

Second, *the need for new models of cooperation to build resilient ecosystems and create positive societal externalities.* Factors such as cyber threat, biodiversity loss, food insecurity or the environmental and health impacts of air pollution are all interconnected and will pose severe challenges that traditional business management education is not fully equipped to overcome. And innovation cannot scale alone. In order for scientific methods and pilot breakthroughs developed in laboratories to meet societal priorities at commercial scale, responsible business practices need to be accompanied by well-funded support and policies that foster public adoption. Portfolio approaches to innovation are set to deliver higher return on economic and societal investment.

Lastly, *the development of governance structures that yield accountability and organizational incentives that align day-to-day business decisions with environmental protection and equitable socio-economic development.* Innovation that can scale starts with a culture of cooperation from within (i.e., with the incumbents).

The 2020–2030 decade will be built on new partnerships and cooperation among organizations and on innovative ways of delivering scientific breakthroughs through open collaboration. In a nutshell, it will be the decade of what I call *pricing change at scale.* These core areas are not fully captured by the traditional lines of defense to mitigate business vulnerabilities by any one organization. They are, in fact, about risk mapping of those vulnerabilities in light of the external impact that response planning efforts are most likely to have on both current and future stakeholders. To become truly resilient, businesses will

DOI: 10.1201/9781003212225-11

need to build accountability for systemic dependencies – the set of interconnected outcomes among core areas – that may not be accounted for in a standard impact assessment.

Where can scientific methods help the most?

THE ROLE OF SOCIAL ENTERPRISES: NURTURING INNOVATION ECOSYSTEMS

There has been an unprecedented pick-up in data innovation (e.g., in sustainability data offerings) stemming from the need to measure non-traditional sources of risks for businesses and translate them into quantifiable, objective metrics to manage and benchmark against. The investors' landscape and the boom in ESG integration practices for financial actors have added to the push for novel sources of data. At the same time, virtually across all sectors of the economy, product innovation has followed. We have seen this in technology and financial services, in consumer products with sustainable features, and in responsible sourcing activities for supply chains. Electrification, energy efficiency applications and breakthroughs in energy technologies add to the ecosystem of innovations. Yet, *system innovation* – the range of organizational processes that help businesses to adapt their talent, operations and relationships to the new set of sustainability challenges and scale to benefit from the growing opportunities – has lagged.

When built on transparency and open architecture opportunities for the purpose of advancing scientific methods, innovation becomes a foundational element for delivering impact outcomes at scale. As organizations are increasingly called to embrace their role as social enterprises, nurturing innovation ecosystems from within or through external cooperation and open networks will underpin their role in building sustainable systems.

When it comes to corporate commitments and funding to promote innovation efforts, what should we prioritize? Anecdotally, it seems that most academics and practitioners tackling the systemic challenges posed by sustainability have devoted their attention to the physical (and possibly, more tangible) risks. These include risks posed by the climate neutrality transition, the social needs of access to resources, the need to bridge the digital divide in the urban/rural context and the need to advance socio-economic development. They have also continued to build a dense repertoire of use cases to help define organizational incentives and governance structures that put accountability first.

Research hubs are redefining lab-scale innovation design to bring sustainability solutions to the marketplace faster. Yet despite the urgent need to tackle the most tangible vulnerabilities our society faces, the transition economy and the set of processes and dynamics that sit at the intersection of the living conditions today and those decades from now are still in the early stages. What I call "transition vulnerabilities" remain unfilled opportunities for social enterprises. These include upskilling people to work with emerging energy technologies, the circularity of materials, or the digital knowledge brought about by on-demand learning or virtual exchange of goods and services.

AN EXAMPLE OF IMPACT INNOVATION AT SCALE

Global payments technologies are likely to become important enablers of effective policymaking in advancing one of the most pressing issues in society today: Climate change.

Since the first carbon-trading scheme was launched in 2005, financial innovation has encouraged both the public and private sectors to exchange international carbon credits. This works in a similar way to how commodities such as energy, metals and agricultural products trade in organized markets. Since 2016, the Paris Climate Agreement is the leading international effort that establishes a framework to limit global temperature rise by lowering greenhouse gas emissions. It includes measurable country-level emission reduction targets as well as mandatory and verifiable reporting of progress.

On the back of the Paris Agreement, active interest in carbon markets has led to the customary use of an internal price for carbon by government entities and private sector enterprises to guide scenario planning across production, distribution and procurement decisions. This was a much needed first step. However, critics of carbon-trading schemes continue to point to its voluntary nature as well as its decentralized jurisdictional oversight and argue that policymaking may never contribute meaningfully to lower carbon economies. Instead, payment technologies may be best positioned to help bridge the gap between short-term economic interests and society's need for a resilient climate.

THE MISSING LINK IN CARBON MARKETS: PAYMENT AGILITY

As reported by the World Bank, carbon-trading schemes are present in over 70 national and subnational governments. They have emerged from a fascinating history built on many attempts to design financial incentives with the purpose of tackling climate change. The term "carbon market" encompasses carbon credits (allowances for carbon emissions under cap-and-trade schemes) as well as carbon offsets. There are two types of benefit here:

- As a tradable unit of carbon in the form of a certificate, *carbon credits* afford the holder the right to emit one metric ton of greenhouse gas). Total available permits are expected to decline over time, and the net effect of this is to provide an incentive for participants to reduce their carbon-emitting activities and become less dependent on a trading plan.

- *Carbon offsets* are also linked to direct investments in projects that are designed and implemented with carbon reduction in mind. These include land and forest preservation, renewable energy and energy retrofit projects, whose processes and outcomes are subject to official verification. In recent years, the increase in popularity of offsets has been primarily due to the additional time they give corporate buyers to redesign their organizational footprint across geographies while minimizing the "internal" cost of carbon-emitting activities.

MULTIYEAR HORIZONS REQUIRE A NEW SOLUTION

It takes years to reach emission reduction targets. Because there is no agile mechanism to assign a fair and comparable financial evaluation of both offsets and credits, there are likely to be fewer transactions in global carbon markets. Technology advances in "payment agility" can address this by making it possible to consistently track carbon emissions in real time, by geography and economic activity, all the way to the underlying emitters.

In addition, establishing a market-driven pricing strategy would make it possible to use traditional pay-as-you-go models. This would be a simpler way for carbon markets to grow to a vast scale and create a measurable record of impact than using traditional commodity exchanges. Pay-as-you-go platforms linked to the internal price of carbon would allow exchange participants to calibrate their willingness to pay to decrease their outstanding carbon balances and move the internal price to an equilibrium versus other counterparties in the exchange.

Early adopters of this practice are likely to include airlines and transportation and shipping companies looking to lower their pollution burden. They could do this by directly establishing a virtual wallet for carbon balances per route and length of travel segment and digitally exchanging their carbon credits with other interested parties.

THE BENEFITS OF A PAYMENTS MODEL FOR CARBON PRICING

From a policymaking perspective, there are two benefits of adopting a payments model to determine carbon price fluctuations:

1. **The establishment of an agile marketplace**: The ability to consistently assign financial value to positive or negative externalities from carbon reduction activities will build trust in the future viability of carbon markets. In fact, what appears to be missing when evaluating the impact of traditional cap-and-trade schemes is a measure of carbon intensity per unit of transacted value. Pay-as-you-go wallet technologies provide the necessary backdrop to build a systemic management of carbon budgets through collaborative efforts among market participants, as opposed to addressing instances of policy failures among the myriad national and international policy programs.

2. **Computational efficiency and digital ledger potential**: Although digital payment authentication is still coping with the low adaptability of existing infrastructure for merchants and financial institutions, it has become easier to envision a digital ledger solution that integrates payments technology and carbon market dynamics. For international efforts such as the Paris Agreement to succeed, significant granularity at the regional, sector and company level is required, in the same way that payment providers follow transacted value. The launch of the Task Force

for Climate-Related Financial Disclosures in 2017 points directly to the importance of tracking and reporting carbon emissions in financial terms by linking carbon emissions to transaction-level activity in the same way that payments technologies are designed to deliver.

The difference between traditional carbon markets and their virtual wallet alternatives is the seamless tracking of carbon removed in aggregate from the system. A virtual wallet is also a more transparent mechanism. It incentivizes participants to disclose their willingness to pay to lower their exposure to carbon risk across their operational or geographic reach. In addition, from a risk management perspective, it would be easier to monitor and recalibrate default versus pre-committed carbon targets set by corporate or governance bodies as a function of the changing range of activities in which they engage.

Cash-for-Carbon: A Look into the (Near) Future

The integration of alternative data (the non-financial data used to support decision-making) through payment mobility apps is introducing a number of use cases, including compelling business-to-business and business-to-customer carbon-tracking solutions. The ability to reduce the carbon footprint of the daily activities of businesses and of individuals as consumers and employees is a key trend to watch. While a variety of economic sectors are likely to augment their environmental and social stewardship through the integration of payments innovations, the biggest benefit is likely to be realized in streamlining supply chains in countries with the highest exposure to adverse carbon-emitting economic policies.

In the case of emerging markets, a growing number of small- and medium-sized enterprises have been the primary advocates of payment solutions to leverage direct access to everyday capital while also creating a direct stream of measurable economic value from emission reductions. As payments solutions are built to address the financial needs of a fully functional circular economy, they are positioned as the most efficient channel to direct and manage carbon sequestration payments, cash-for-carbon municipal programs and congestion-charging schemes. The integration of mobility payments to address the need to transition to a low-carbon economy is set to create the next generation of pay-for-good business models.

IMPACT-ORIENTED SUSTAINABILITY TARGETS: FROM ADOPTION TO IMPLEMENTATION

Earlier in my discussion of the role played by organizational learning in stirring the sustainability journey of businesses, of trust and transparency as impact variables in setting environmental and social commitments, I highlighted the rise of open innovation in sustainable systems. When businesses operate in economic and social systems that allow open innovation to help achieve impact-oriented targets, it is necessary to maintain a multistakeholder dialogue to foster a productive collaboration network. In turn, this means that active engagement with all parties involved is prioritized. In fact, as the strategic growth roadmap of businesses puts socio-economic and environmental impact at its core, implementation of commitments involving the human and natural capital that sustain a business can translate more effectively in a step-by-step approach to organizational learning and a dynamic adoption of sustainability commitments.

The experience of many of the most well-resourced companies in the world – in terms of teams, time and financial capital – shows that the choices of publicly committing to a net zero carbon economy, closing gender pay gaps or setting equitable representation in their workforce require them to engage at scale within and outside their traditional value chain. It forces them to become advocates of the change their commitments aim at achieving, employing trust and transparency every step of the way. It also means employing both a sustainability lens and an innovation lens in collaborating across ecosystems to bring the future state of sustainable living, sustainable working, sustainable producing and sustainable consuming to the center of their business model evolution.

One example of the connection between innovation and sustainable transitions is presented in the Sustainability Accounting Standards Board (SASB) Materiality Map®. The application of materiality considerations by industry to define the sustainability dimensions faced by companies helps a wide range of stakeholders to engage on the effectiveness of accounting and the long-term value of reporting in alignment with the SASB standards. "Business Model and Innovation" is one of the five sustainability umbrella dimensions identified by the SASB, along with the Environment, Social Capital, Human Capital, and Leadership and Governance. In SASB's own definition:

> This dimension addresses the integration of environmental, human, and social issues in a company's value-creation process, including resource recovery and other innovations in the production process; as well as in product innovation, including efficiency and responsibility in the design, use phase, and disposal of products. (SASB 2017, 3)

Applying the Sustainability Lens to Innovation Ecosystems

The companies and leaders involved in defining and delivering organizational impact targets may be directly stepping into an innovation ecosystem for the first time. More often than not, moving from adoption of sustainability commitments to implementation requires a combination of both *data innovation* and *system innovation*.

In this context, I define *data innovation* as the family of processes that encompasses research, prioritization, modeling and integration of all the unique data sources that best describe the opportunities and challenges posed by sustainability dimensions – either to the business itself or its ecosystem within a specific timeframe (most likely the medium to longer term).

System innovation is the set of scenarios that best describes the transition of an organizational setting, business model, or industry as a result of sustainability dynamics. An example is the transition of electric utilities or transport to renewable sources of energy.

Since the launch of the UN Sustainable Development Goals in 2015, academics and industry practitioners have been overwhelmed by data innovation to define and maintain the sustainability and impact assessment of businesses, sectors and countries. Whether in the form of ratings, rankings, scores or open data feeds, sustainability has become an information marketplace for both incumbents and innovators, driven by advances in cloud computing, artificial intelligence and machine learning applications, to name just a few. While there is no shortage of data feeds, the poor quality, narrow scope and depth of coverage of data sources have been occupying most of the practitioners' time as organizations face the first challenge in the adoption of sustainability commitments: Either voluntary or regulatory reporting of reliable and verifiable impact metrics.

What has been lagging is *system innovation* for lasting adoption and implementation of sustainability commitments through scenario planning. System innovation for sustainability usually involves a number of actors (e.g., scientists, academics, innovators, private and public sector representatives, civil society). Through longitudinal studies and qualitative research for smart city projects, Oskam et al. (2021) highlight how participants in innovation ecosystems are faced with three dynamics in their co-creation of value:

1. the dynamic of value creation vs. value capture
2. the dynamic of collective vs. individual value
3. the dynamic of gaining vs. losing value (e.g., whether value created is fairly distributed and captured among participants)

According to the researchers, these dynamics originate as the participants aim at identifying the value proposition of the ecosystem. How do they reach an agreement on the degree of environmental, social and economic benefit to create through their collaborative effort? How does the value co-created get redistributed among participants? How does it balance off each participant's interest?

This collective discovery may slow down the progress and make multilateral collaborations cumbersome unless participants engage in either "collective orchestration" or "continuous search." The former allows the ecosystem's actors to be changed, making it possible for participants to enter and exit in a learn-and-experiment setting as the ecosystem evolves over time. The latter stems from action-based learning and incorporates new ways to create value and seek new partners as the outcomes of experimentation provide feedback. Both are pathways to value creation that integrate economic viability and societal value.

APPLYING THE INNOVATION LENS TO SUSTAINABILITY TRANSITIONS

The transition of a business to a sustainable model of economic and societal growth ("sustainability transition") requires both data innovation and system innovation. At the 2019 IEEE International Symposium on Innovation and Entrepreneurship, Prof. Michael Zhang from Nottingham Business School traced the technological innovation that has been revolutionizing the global automotive sector to comply with a low carbon economy – an innovation driven by policy interventions, not open collaboration networks. Prof. Zhang argues that innovation ecosystems thrive because of the multilateral web of stakeholders involved in effecting mutually beneficial change alongside a common proposition of sustainable value. For example, smart mobility solutions have not been pioneered by leading vehicle manufacturers – the ones most affected by international regulatory pressures to curb air pollution and environmental degradation. On the contrary, it is the intermediate users – those who use vehicles to distribute services and products – who have defined the path to sustainable innovation for green transport.

At this point, it will come as no surprise that a big part of the challenge is indeed developing open collaborative networks. Those networks must be built in an environment of trust and allow continuous evolution of the value preposition(s) surrounding both common sustainability goals and those set by individual participants. A balanced, transparent engagement with the actors that may enter and exit the ecosystem during the transition is required. This allows for experimentation to generate individual and group learning milestones and reassess the feasibility of the targeted impact. Applying the innovation lens to the sustainability transition of a business requires a cultural openness to peer dynamics. This in turn allows group dynamics to be balanced with the participant's individual, trial-and-error-based learning.

In my contribution to *Applying Neuroscience to Business Practice* (Dos Santos 2017), I discuss how social learning dynamics are in fact likely to replace expected utility frameworks (e.g., the maximization of individual benefit) in purely financial decision-making. After years of participating in leading international efforts that promote sustainable value creation in financial markets, I argue that sustainability transitions of businesses and sectors that employ innovation ecosystems to thrive *in a system setting* can be an invaluable complement to the application of data innovation alone.

A "less is more" approach helps to establish which qualitative or quantitative observations (data) are the most relevant and decision useful as descriptive marks on the sustainability transition. Examples include the carbon intensity of a business relative to its peers in a net zero carbon economy commitment to a baseline year, and gender representation, employee turnover and average pay in a diversity and inclusion commitment for the business to promote equality. However, by adopting the key features of trust (vulnerability, risk and expectation) as indicators of the adaptive learning needed in the transition, choices that are optimal system-wise, as opposed to being optimal for the individual participants, can be prioritized.

Neural brain systems that enable learning can help bridge the gap between individual experiential learning and group dynamics. They also provide a strong theoretical foundation for the use of system innovation to embed sustainability learning as part of an organization's DNA, creating economic and societal capital as part of a new innovation-driven engine of growth. (Refer to Technical Note for an application of the innovation lens to sustainability transitions involving circular economy outcomes.)

THE LEARNING JOURNEY – THE NEXT DECADE

- The 2020–2030 decade is a pivotal time for businesses to redefine their role in driving economic and societal wellbeing. The rapid flow of externalities that challenge our global context has unveiled a series of systemic dependencies from non-traditional environmental, social and governance sources that organizations need to effectively quantify to manage.
- The booming landscape for sustainability data and product innovation continues to redefine common measures against which standards of responsible business and financial management are benchmarked. System innovation – the wide range of processes that influence the effectiveness of evolving business models to deliver large-scale sustainability objectives – has lagged behind.
- While multiyear horizons require a novel set of management solutions, the ability to leverage open technological innovation across sectors will be key to develop a thriving business ecosystem and enable sustainable transitions of business activities where multilateral cooperation efforts provide pathways for co-creation of socio-economic value and mutually beneficial change.
- Circular economy models introduce regenerative consumption-production systems and allow for a dynamic assessment of business vulnerabilities as well as value-capture opportunities during a sustainable transition.

TECHNICAL NOTE – CIRCULARITY MODELS FOR SUSTAINABILITY TRANSITIONS

The term "circularity" is often used to describe the set of process and product innovations that address the reuse cycle of natural resources. They do this in three ways: extending the usable life of materials and products; minimizing environmental degradation as a result of optimized materials' use and products' consumption; and helping to regenerate the natural systems that produce natural resources. Therefore, "circular economy" refers to an economic system that relies on widespread adoption of circular processes and product innovations. This breaks the vicious cycle in which finite resources are consumed for production purposes and economic growth.

According to the annual Global Circularity Gap Report, which has been published during the World Economic Forum in Davos since 2018 as part of the Platform for Accelerating the Circular Economy (PACE), it is estimated that as of 2020 less than 9% of economic growth globally relies on circularity principles. Notwithstanding the global need to close the gap, emerging examples of circular models adopted by businesses at the micro level offer a rare combination of data, process and system innovations ("sustainability transitions"). Together, these can address the consumption of primary resources while also supporting the development of production and manufacturing activities that use regenerative solutions.

The sustainability transitions presented in Table 10.1 offer valuable examples. They draw from the broader case study of the Ellen MacArthur Foundation. The Foundation's pioneering method of measuring a company's circularity performance (Circulytics®) offers a series of indicators (*Material Circularity Indicators – MCI*) and presents the assessment result in a scorecard template. While the Circulytics® method is quite data-intensive for businesses, it relies on the assumption that company level MCIs can be derived as a proxy of the aggregate of individual product's circularity statistics and the flow of materials volume involved in the manufacturing of each product and its offering in the marketplace. What is likely to be lost in the aggregation stage are the changing real-world conditions the business is adapting to and how well synchronized the changing business model is with the stakeholder community it touches. By adding a set of *complementary risk and impact indicators* to the aggregation of individual product assessments, a circularity framework built around the MCI model helps to identify the *material* circularity effects on the business or its stakeholders. Therefore, business priorities for crafting a sustainability transition roadmap with measurable outcomes from system innovations can be defined.

Let us look at an example, using the case of sustainability transition for the consumer products segment in Table 10.1. Here, a *complementary impact indicator* would be the assessment of the effect of the *consumer education* programs on *product recyclability*. It is listed as a way of evaluating a value-capture opportunity for the company, as it potentially drives higher awareness and develops a perception of higher brand value in the medium term. It also exposes the organization to value-creation opportunities in the alignment with other best-in-class organizations – within its own end markets or outside. Also, any *expectations of scarcity and toxicity* of materials – within the standard procurement cycle or as a result of sudden supply chain risks – or *price volatility of materials* would constitute *complementary risk indicators* (product-level and/or company-wide).

Bianchini et al. (2019) highlight the importance of breaking down the barriers to the actionable implementation of circular business models with a new visualization tool: the "Circular Business Model" (CBM). CBM promises to combine the measurement of materials/resource flows plus other indicators of environmental and societal impact with a focus on the entire value chain of a business, as opposed to as the aggregation of its products' materials footprint and its vulnerabilities in primary end markets.

TABLE 10.1
Examples of Sustainability Transitions

Ecosystem Characteristics	Value Proposition	Innovation Type	Potential Value Creation	Potential Value Capture
Sector: Food and Beverage • Cross-functional innovation teams • Start-up founders participate in pilot program	• Reduction of single use packaging	• Product innovation (materials) • Process innovation (sourcing cycle)	• Enabling circular procurement process • Increased use of recycled materials in products	• Launch of collaboration platform across sector to reduce single use packaging and waste • Potential for direct adoption of venture pilot programs
Sector: Consumer Products • Set up of research lab with academic partners • Cross-sector partnerships (waste management sector)	• Enhance plastic bottles' recyclability • Ramp up waste collection and remake cycles	• Product innovation (materials) • Process innovation (waste collection cycle)	• Increased use of recycled materials in products • Brand equity from cross-sector strategy alignment	• Creation of circularity materials flow (plastics) through public-private collaboration • Community programs to educate consumer on product recyclability
Sector: Automotive • Functionally focused (maintenance; R&D) • Fully in-house – stage 1; no external actors involved (regulatory driven) • Stage 2 multi-stakeholder collaboration	• Establish short-loop recycling of plastics • Optimized materials sorting capabilities	• Process innovation (materials waste; customer bonus for remanufactured components)	• Innovation partnership with local policymakers • Potential for process R&D on enhanced materials sorting capability	• Potential for closed-loop recycling method to other raw materials • Potential for stronger collaboration in local markets (workforce skill development)

(Continued)

TABLE 10.1 (*Continued*)
Examples of Sustainability Transitions

Ecosystem Characteristics	Value Proposition	Innovation Type	Potential Value Creation	Potential Value Capture
Sector: Electronics **Manufacturing & Servicing** • Grassroots global engineering community (subject matter expertise) working in open network on key innovation • Set up of innovation Center of Excellence	• Increased automation across value chain (e.g., manufacturing, delivery and after-sale support)	• Product innovation (tracking software for reverse logistics) • Process innovation (automation in repair & refurbish cycle) • System innovation (after-sale support of automated user testing)	• Increased customer satisfaction and brand awareness on quality, user-friendly automated DIY after-sales support	• Increased engagement with consumer communities on reverse logistics programs • Increased consumer education on at-home servicing and automated testing

Source: Author's research and intuition; example drawn from the Ellen MacArthur Foundation, Circulytics®.

THE ROLE OF MATERIALITY IN ADOPTING CIRCULARITY INDICATORS

A materiality lens, such as the one outlined in the Sustainability Accounting Standards Board (SASB) Materiality Map®, can be overlayed by industry sector to provide a powerful implementation tool to streamline the intense data collection exercise of building an aggregate circularity model for organizations of different sizes or product footprints. In fact, designing circularity indicators (such as MCIs) with a financial and economic materiality scope in mind will help embrace cross-functional accountability and define the baseline for acceptable outcomes. It also brings to life an internal organizational compass for closing circularity gaps as they emerge. Intuitively, materiality unveils opportunities for system innovation early in the process.

When choosing circularity indicators that are aligned with both value creation and value capture associated with a sustainable transition of the consumer goods industry (e.g., building products and furnishings), the SASB Materiality Map® points to four critical dimensions. These dimensions are material for the purpose of disclosing sustainability information that is decision useful to stakeholders of the business:

1. Energy management in manufacturing
2. Management of chemicals in products
3. Product lifecycle environmental impacts
4. Wood supply chain management (where applicable)

The SASB dimensions can be overlayed with traditional flows of materials and primary resources and the innovation type involved in making circularity solutions (e.g., product, process, system) economically self-sustaining and scalable

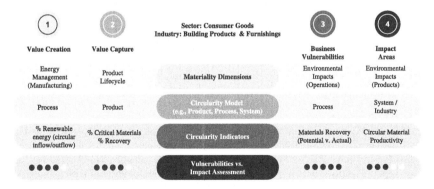

FIGURE 10.1 Example of circular business model dynamics – consumer goods. Author's intuition. Materiality Dimensions: sample dimensions from industry-level SASB Materiality Map®; Circularity Model: sample categorization by type; Circularity Indicators: sample indicators from the WBCSD (World Business Council for Sustainable Development); Business Vulnerabilities and Impact Areas: sample assessment from author. For illustrative purpose only.

as enablers of quantifiable positive impact. This offers a powerful representation of existing business vulnerabilities and reveals which potential areas of impact should be prioritized.

It is worthwhile recalling the role of the Informal Organization, as discussed in Chapter 8. In the layout and selection of circularity indicators that best align a company's sustainability journey with its existing operational capabilities, its innovation needs and, potentially, its business ambitions, modeling system interactions is key to adoption. Successful impact assessments – those that deliver sustainable business outcomes and help map where and when system innovations are best suited to occur – are iterative, multistakeholder processes rooted in collective discovery and value co-creation.

REFERENCES

Abreu, A. 2021. "Innovation Ecosystems: A Sustainability Perspective." *Sustainability* 13 (4): 1675. doi:10.3390/su13041675.

Bianchini, A., J. Rossi, and M. Pellegrini. 2019. "Overcoming Barriers of Circular Economy Implementation Through a New Visualization Tool for Circular Business Models." *Sustainability* 11: 6614. doi:10.3390/su11236614.

Dos Santos, M. A. 2017. *Applying Neuroscience to Business Practice*. Hershey, PA: Hershey.

Ellen MacArthur Foundation. 2015. "Circularity Indicators: An Approach to Measuring Circularity." Project Overview. https://www.ellenmacarthurfoundation.org/assets/downloads/insight/Circularity-Indicators_Project-Overview_May2015.pdf.

Falsarone, A. 2017. "Neuroscience Applications in Financial Markets: A Practitioner's Perspective." In *Applications of Neuroscience: Breakthroughs in Research and Practice*, edited by Dos Santos, M. A., 115–141. doi:10.4018/978-1-5225-1028-4.

Falsarone, A. 2019. "How Payment Agility and Cash-for-Carbon Can Solve a Global Problem for Mankind." In *The PayTech Book: The Payment Technology Handbook for Investors, Entrepreneurs and Fintech Visionaries*, edited by S. Chishti, T. Craddock and R. Courtneidge, 158–161. Wiley online library. doi:10.1002/9781119551973.ch47.

Haigh, L., M. de Wit, C. A. Colloricchio, and J. Hoogzaad. 2021. "Circularity Gap Report." PACE. https://drive.google.com/file/d/1MP7EhRU-N8n1S3zpzqlshNWxqFR2hznd/edit.

Oskam I., B. Bossink, and A.P. de Man. 2021. "Valuing Value in Innovation Ecosystems: How Cross-Sector Actors Overcome Tensions in Collaborative Sustainable Business Model Development." *Business & Society* 60 (5): 1059–1091. doi:10.1177/0007650320907145.

SASB. 2017. "Conceptual Framework." February 2017. https://www.sasb.org/wp-content/uploads/2017/02/SASB-Conceptual-Framework.pdf.

Swiss Re Institute. 2021. "The Economics of Climate Change: No Action Not an Option." April 2021. https://www.swissre.com/institute/research/topics-and-risk-dialogues/climate-and-natural-catastrophe-risk/expertise-publication-economics-of-climate-change.html.

Zhang, M. 2019. "Innovation Ecosystems for Sustainability Transition: From Policy Intervention to Stakeholder Coalition." *2019 IEEE International Symposium on Innovation and Entrepreneurship (TEMS-ISIE)*: 1–10. doi:10.1109/TEMS-ISIE46312.2019.9074176.

Conclusion

In its Sixth Assessment Report on Climate Science released during the summer of 2021, the Intergovernmental Panel on Climate Change unambiguously brought forward scientific evidence of Earth's temperature warming as a human-caused phenomenon. Adaptation and mitigation efforts are being carried out by businesses as a result of the dire environmental degradation we are confronting. But to be effective, those efforts must take today's sustainability pledges and elevate them to programmatic and intentional objectives that realign business priorities with societal priorities. Whether it is building environmental resilience, addressing climate justice, or building inclusive knowledge platforms, the impact challenge is a call for sustainable business model transitions.

If I have done my job right in bringing the concepts and the lived experiences behind this book to life, you are now able to recognize that impact is a multifaceted matter, regardless of the angle you start with. Natural Capital, among others, is a new field of research that highlights the interdisciplinary nature of the toolkit needed to maximize the regenerative power of our planet, balance our societal needs and define a new economic backdrop for businesses to flourish. You should also be walking away with a handful of key tenets and reflection steps to inspire the beginning of your own contribution to solve the impact challenge of this decade.

Yet, in the age of ESG leaders versus laggards, of screening according to improving versus lagging sustainability indicators, it is a good reminder to everyone that a new industry was born almost overnight on the premise of mastering novel indicators and turning the clock on progress.

There are few perils to this lens. It is frequently used for scrutiny, not as a gauge of progress and constructive dialogue. It gives minimal incentive to co-create value among peers. It depicts organizations that progress on their own, not in tandem as they advance best practices.

This is something counterintuitive for all of the professionals who, like me, have had the good fortune of experiencing how collaborative and open to co-creation of value and sharing of best practices the field of ESG can be. The science behind sustainability comprises a multitude of system-wide dynamics that require the sharing of knowledge and resources to counterbalance them as opposed to a winner/loser mindset. When it comes to sustainability, if one loses, we all lose.

I recognize a few benefits of the ESG leaders/laggards comparisons when they are deployed to build awareness and reward the hard work of many within the organizations that emerge as leaders. What the leaders/laggards mindset misses is how to go from a laggard to a leader. How one could fall behind as a leader if sustainability best practices remain encapsulated as a set of policies for display on a website or sustainability report – a set of policies to stay compliant with the latest regulatory "safeguards" or "minimums" as opposed to an ongoing practice of leadership introspection.

My wish for you, the reader, is to make sustainability principles alive as part of your role, day in and out, regardless of having "ESG" or "impact" in your job description. I invite you to start the journey by comparing lessons learned in this book with actionable insights and share your reflections with your colleagues and the industry networks you contribute to. Lastly, I encourage you to manage your professional and personal aspirations in alignment with the sustainable transitions we are facing in the decade of ESG. Push beyond blueprints, or tailor one that fits your own ambition.

To the naysayers: Please help spread the word, the reasoning behind why you believe there may be flaws and inconsistencies in the approach presented in this book and what you would suggest otherwise. By doing so, you are fostering an open dialogue in a purposeful way on possibly the most pressing matters of our lifetime.

Index

Note: **Bold** page numbers refer to tables; *italic* page numbers refer to figures and page numbers followed by "n" denote endnotes.

Printed in the United States
by Baker & Taylor Publisher Services